HISTORY

OF

BRITAIN

IN

MAPS

Published by Collins
An imprint of HarperCollins Publishers
Westerhill Road
Bishopbriggs
Glasgow G64 2QT
www.harpercollins.co.uk

First published 2017

A catalogue record for this book is available from the British Library

ISBN 978-0-00-825834-4

10 9 8 7 6 5 4 3

Printed in China

MIX
Paper from
responsible sources
FSC™ C007454

This book is produced from independently certified FSC™ paper
to ensure responsible forest management.

If you would like to comment on any aspect of this book, please contact us at the above address or online.
e-mail: collins.reference@harpercollins.co.uk

 facebook.com/collinsref

@collins_ref

HISTORY
OF
BRITAIN
IN
MAPS

Philip Parker

Contents

" ... In that Empire, the Art of Cartography attained such Perfection that the map of a single Province occupied the entirety of a City, and the map of the Empire, the entirety of a Province. In time, those Unconscionable Maps no longer satisfied, and the Cartographers Guilds struck a Map of the Empire whose size was that of the Empire, and which coincided point for point with it."

Jorge Luis Borges, On Exactitude in Science, 1946

Introduction

Maps are a very ancient way of making sense of our world. From scratches on rocks made in Neolithic times and ancient Mesopotamian cuneiform tablets to the globe-spanning digital databases of the twenty-first century, they have translated a visual account of a landscape or of a nation into an image that it would take many thousands of words to equal.

Maps are an imperfect mirror of reality. They may approach, but never reach Borges's mythic map where they become the monstrous equal of the area they portray. Yet however accurate they become, maps reflect so much more than merely a growing knowledge of the countries in which their compilers lived or a steadily increasing level of technical skill in transferring this cartographically onto the page. The subjects chosen, the information included (or left out) and the manner of presentation are all influenced in part by the preoccupations of the age in which they were drawn. The priorities of the author of the thirteenth-century Hereford *Mappa Mundi*, with Jerusalem at its centre, and Britain decidedly at its periphery, were very different from those of the maker of the map of the 1841 British census, for whom mapping man, and not God, was the central concern.

History of Britain in Maps recounts the story of England, Wales, Scotland and Ireland by reflecting on what those maps can tell us about the motives of the mapmakers and the history of the eras in which they lived. The 1571 map by William Bowles of Elizabeth I's progress (or royal visit) to Norfolk is at one level a prosaic administrative document used for planning the queen's provincial trip, and yet at another it is a sign of the need of early modern monarchs

An extract from The Hereford *Mappa Mundi* c.1290 showing the British Isles.

to reinforce the mystique of their rule by very solid appearances before their more aristocratic subjects. John Elphinstone's 1745 map of Scotland represents a refinement in the cartographic portrayal of Scotland, but it also sits squarely in a period when the nation was in turmoil, racked by two unsuccessful uprisings to restore the deposed Stuart dynasty to the British throne, and was issued in the year before the defeat of the second, under Bonnie Prince Charlie, would lead to the destruction of the age-old Highland way of life.

The maps in this volume span nearly two thousand years, from the Rudge Cup, a bronze Roman vessel which bears a representation of the line of Hadrian's Wall, to a map showing the distribution of voting in the 2016 referendum on whether Britain should leave the European Union. In the time in between, cartographers have mapped routes, property disputes, defensive systems, battles, mineral resources, railways and canal networks, the weather and even the progress of a cholera epidemic; anything in short where the placement of a line, a symbol or an area of shading could display information effectively, tell a story or promote a message.

Even for the first half of the period when, after a brief spark under the Romans, there is no contemporary mapping made in Britain until the eleventh century, later scholars and antiquarians so much understood the value of the map that they created them to help illustrate their own distant past. Hence, the thirteenth-century chronicler Matthew Paris's maps of the pre-Roman roads of Britain and the Heptarchy (or seven kingdoms) of Anglo-Saxon England are fascinating both for the medieval notion of England's very early history which they encapsulate, and as a reminder that the political shape of Britain was once very different from the simple evolution that a concentration on events after the Norman Conquest might suggest.

Among the many concerns that mapmakers have had through the ages, two are most prominent – to chart the realm accurately (through topographical maps, road itineraries or plans of transportation systems) and

to document its defence or the struggles to maintain its integrity from attack by others. From the first we can learn of physical changes, of how, for example, the proposal to build a canal from Stockport to Darlington reflected the insatiable hunger of Britain's factories for raw materials, in particular coal, during the early stages of the Industrial Revolution; or how a map of the centre of late-medieval Bristol sheds light on that city's central place in the first wave of British voyages to North America in the 1490s. The imperatives of war are more obvious. Each invasion scare – and there

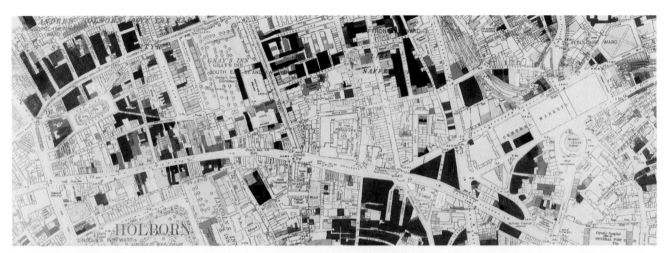

An extract of Robert Ricard's 1479 *Plan of Bristol* (OPPOSITE) and a map of bomb damage to London during the 1940 Blitz (ABOVE).

were many between 1066 and 1940 – called for more accurate maps of Britain's coastlines and work on better defensive fortifications which in turn needed maps and plans to record them.

And when diplomacy failed – or was not even tried – the mapmakers stepped in to chart the course of the periodic conflicts between England and Scotland or of the English Civil War itself, where a plan of the siege of Newark in 1646 is all the more poignant for knowing that right at the end of it Charles I surrendered to the Scots, losing control of his own person and beginning the slow march to the scaffold which ended with his execution on 30 January 1649. Finally, when fear of invasion turned to reality, maps are some of the most eloquent witnesses to the destruction that resulted; in maps cataloguing the damage caused by the Blitz

against Britain's towns and cities from 1940 and in displaying the route of German bombers during the far less well-known German aerial assault during the First World War.

Every map tells a story. From the smallest – dealing with a property dispute in medieval Yorkshire – to the largest – portraying the whole of the United Kingdom and the density of population after the 1841 census – they are amongst the most eloquent forms of historical narrative. The many maps in this book, with their varied perspectives, motives, forms of execution and differing geographical frames, each tell their own tale. Together they shed refreshing new light on the wonderful kaleidoscope of patterns that is the history of Britain.

PRE-ROMAN PAVED ROADS,
Matthew Paris

The long story of human settlement in Britain before the Roman conquest of AD 43 cannot be told through documents or maps, as none survive (if they ever existed). However, the work of Matthew Paris, a thirteenth-century monk from St Albans, who created a map showing what he imagined to be the pre-Roman road network, gives a glimpse of how this period of British history was seen in the Middle Ages.

Although a Benedictine monk, Paris was well-connected, being personally acquainted with both Henry III and his brother Richard, Earl of Cornwall, and his access to the extensive monastic library of St Albans ensured a deep knowledge of the English historical traditions of the time. On this map, included in the *Liber Additamentorum* ('The Book of Additions'), a supplement to his principal chronicle of English history, Paris lays out the four routes which were traditionally believed to represent a system more ancient even than the Roman road network.

Paris gives Fosse Way, Icknield Way, Watling Street and Ermine Street their Latin names and has them incongruously intersecting at Dunstable (not known ever to have been such an important transport hub). He links them to the tradition begun in the 1130s by Geoffrey of Monmouth in his *History of the Kings of Britain* that the roads, of very ancient construction, had been renovated by Belinus, king of southern England, a reputed descendent of Brutus, a refugee from the Greek capture of Troy at the end of the Trojan War.

Although in many ways fanciful, there was some basis for Paris's map, as Roman roads such as the Icknield Way (from Norfolk to Wiltshire) and Watling Street (from Dover to St Albans) were built along the lines of ancient pathways which dated back at least to the beginning of the Iron Age in Britain around 800 BC. That some of these were not simple grass tracks is indicated by the discovery of an Iron Age metalled road unearthed at Bayston Hill in Shropshire in 2011, which came as a surprise to archaeologists who had assumed paved roads were a Roman innovation.

Britain became more densely populated at this time, and there is evidence that competition over resources was leading to a more militarized society, as hill forts – defensible high-points, often with earthwork ramparts and palisades – spread throughout southern and central England. Some of these, such as Danebury in Hampshire, were as large as 10 ha (25 acres) in area, with groups of circular houses clustering inside. By the beginning of the first century BC, some had grown to become almost proto-urban settlements (known as *oppida*) which acted as royal centres for tribal grouping such as the Catuvellauni (north of the Thames), the Atrebates (in Sussex and Hampshire) and the Iceni in East Anglia.

High-value goods from the Mediterranean, including wine and olive oil, were imported to capitals such as Verulamium (St Albans) – the main settlement of the Catuvellauni – being landed at Hengistbury Head in Dorset and other ports and from there transported overland. The economic links between the British kingdoms and their counterparts in Gaul grew stronger (extending to the first minting in Britain of coins, which bear some of the earliest personal names we have, such as Tasciovanus, the Catuvellaunian king, and his son, Cunobelinus.

Yet the strengthening of these ties, as the latest luxuries were hauled along Matthew Paris's roads to pander to the tastes of British kings, nobles and warriors, also made those elites vulnerable. For as Julius Caesar's Roman legions fanned out northwards across Gaul from 58 BC, they became aware of the possibilities for glory and enrichment presented by the island to its north. Three years later, the first Roman invasion of Britain took place and ultimately those trackways would become Roman roads and the sinews which bound the island close to its new imperial masters.

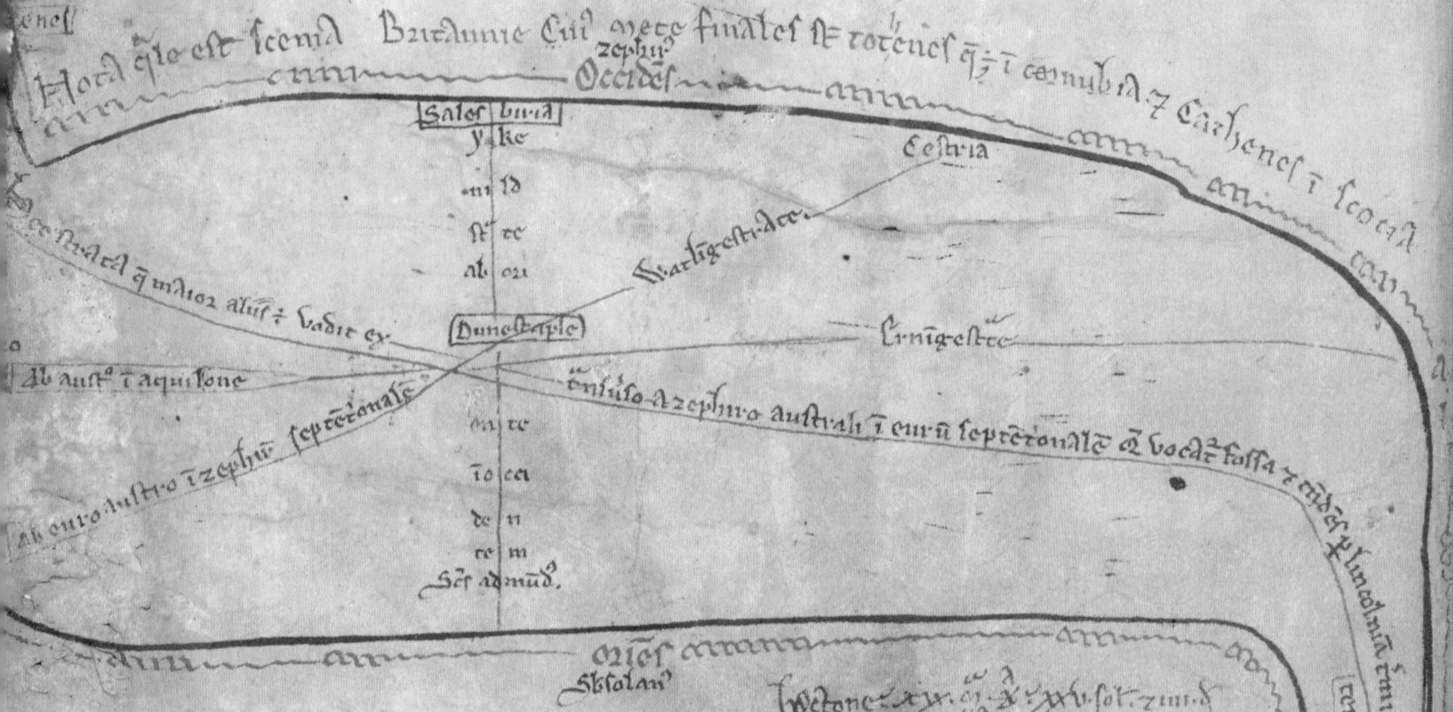

Hoc̄ q̄o est scoc̄ia Britanie cui̅ mē fin̅ates s̄r toteñes q̄ ĩ cō̅nub ĩa 7 cathenes ĩ scoc̄ia

Occidens

zephir̄

Sales buria
y ko
oñ ld
st te
at oñ

Cestria

Wartigestr̄ dec̄

Ec̄ finac̄a q̄ maior aliu̅t̄ vadit ex

Dunestaple

Ernigestre

Ab aust̄ ĩ aquilone

infuso a zephiro australi ĩ eurū septētonalē q̄ vocāt fossa 7 uad̄ 3

Ab euro austro ĩ zephir̄ septētonale

f lincolnia sernabāc̄ia

en te
to ea
de n
re m
Sc̄s admud.

renes

Oriens

Sc̄s olan̅

hetone x̅i̅ m̅ x̅ x̅v̅ sol̅ 7 iiii d̅
lauñ ex l̅ m̅ x̅ v̅ ii ñre
crestighir̅ q̅ii̅ deam̅ x̅ ii̅
brelurne c̅ m̅ l̅ decim̅ x̅ q̅
Poteshue iii m̅ x̅ v̅ sol̅ 7 iiii d̅
tirefeld x̅ii m̅ dec̅ i̅ m̅ 7 x̅x̅x̅v̅ d̅
ec̄a oiu̅ f̅c̅x de subire x̅viii m̅ d̅ x̅v̅ sat̅ 7 A

f Sum̅a decime pr̅ioum̅ ex diocef̅
de ec̄a de dachet ii m̅ 7 dim̅ x̅ i̅ii sol̅ 7 v̅iii d̅ ×
de wuburne x̅v̅ sol̅ x̅ x̅viii d̅

Sum̅a decime elemofin̅arū
de ec̄a de ellyeuñ 7 d minutis decim̅
fiu̅t̄ ex dioc̄ sc̄i albi lx 7 x̅ sol̅ 7 v̅ d̅
f Sum̅a totā fup̅ac̅ x̅lv̅ m̅ 7 lc̅ii
f Sum̅a decime ec̄c̅ ii̅ f dioc̄ q̅ id̅

Sum̅a
xl m̅ xxxu d̅
viii d̅

due ded̅ genic̄
Bnd̄ci deu ilc̅ cui̅ 7 h̅u̅

lic̄ ilr̅
alb̅ cui̅
q̅ci

fer dp̅u

CAESAR'S CAMP AT THE BRILL, William Stukeley

Although the Roman Empire had a long tradition of surveying, both civil and military, very few traces of Roman maps have come down to us. The 450 years of Roman involvement in Britain have left even less in the way of cartography, with antiquarian reconstructions being one of our few routes into the world of Roman maps.

There were clearly large-scale Roman maps – the Emperor Augustus (27 BC – AD 14) is said to have commissioned one of the entire known world – but of these only fragments and later reconstructions of itineraries survived the vicissitudes of the empire's collapse. Into this void stepped scholars and antiquarians, who from the sixteenth century embarked on a process of recording and recovering what they could of Britain's Roman past. William Stukeley's reconstruction of Julius Caesar's camp at the Brill formed part of this programme, purporting to show a marching camp built by one of Caesar's legions in what is now north London during his invasion of Britain in 54 BC.

Stukeley was an Anglican clergyman from Lincolnshire who embarked on a series of wide-ranging tours of Britain between 1710 and 1725, visiting ancient structures from Stonehenge to Hadrian's Wall. His passion for archaeology led him to make detailed drawings and plans of these. At the Brill, on the banks of London's ancient Fleet River (which flows from Hampstead to the Thames, now in underground sewers) near Old St Pancras Church (marked as Pancras Church on the map), he believed he had identified a series of bumps in the ground as the remains of a Roman legionary marching camp. The coloured plan he drew shows the division of the camp into a neatly ordered grid with areas for the various categories of troops; the *hastati*, the younger, inexperienced men, who fought at the front, the *principes*, or more seasoned warriors, and the *triarii*, or veteran soldiers, who were committed to the fray only as a last resort. In the centre is the *praetorium*, or headquarters of the general, next to it the *quaestorium*, or camp of the quarter-master, and around the edge of the encampment a rampart punctuated by four gates. In his accompanying text he speculated on how Caesar may have met there with Mandubracius, a British prince who collaborated with the Roman invaders.

Although it was Stukeley's intention 'to preserve, as well as I can, the memory of such things as I saw; which, added to what future times will discover, will revive the Roman glory among us', his plan of the Brill was fanciful in several respects. The divisions of the legion he shows reflect the situation before a military reorganization some forty years earlier under Gaius Marius (which divided each legion into ten cohorts, and did away with the old distinction between *triarii* and *hastati*), while no one after Stukeley was ever able to identify the Roman remains which he claimed had stood there.

Julius Caesar did, though, pass through the London area during the second of his invasions of Britain, quick sallies across the English Channel which provided occupation for his troops at a time when his recent conquests in Gaul seemed quiescent. The first attack, in 55 BC, had almost ended in disaster when the transport ships of the relatively small invasion force (of two legions) were wrecked off the Kent coast. The following year Caesar returned with five legions and 2,000 cavalry and, after a contested crossing of the Medway, chased the British leader Cassivelaunus northwards as far as St Albans before – having taken tribute and hostages – crossing back to Gaul, never to return.

What significant Roman remains there are in Britain date from the era following a later invasion ordered by the Emperor Claudius in AD 43, which had, within forty years, subdued all of Britain south of the Scottish Highlands. Legionary marching camps, and more permanent stone forts, studded the landscape as far north as Inchtuthil in Perthshire. Their basic structure (of a grid of tents or barracks, divided by roads and with set positions for the senior officers' quarters) reflect the layout which Stukeley had given his imagined reconstruction. Whilst not a genuine Roman map, therefore, it is certainly a precious insight into how the scholars of past centuries viewed their Roman heritage.

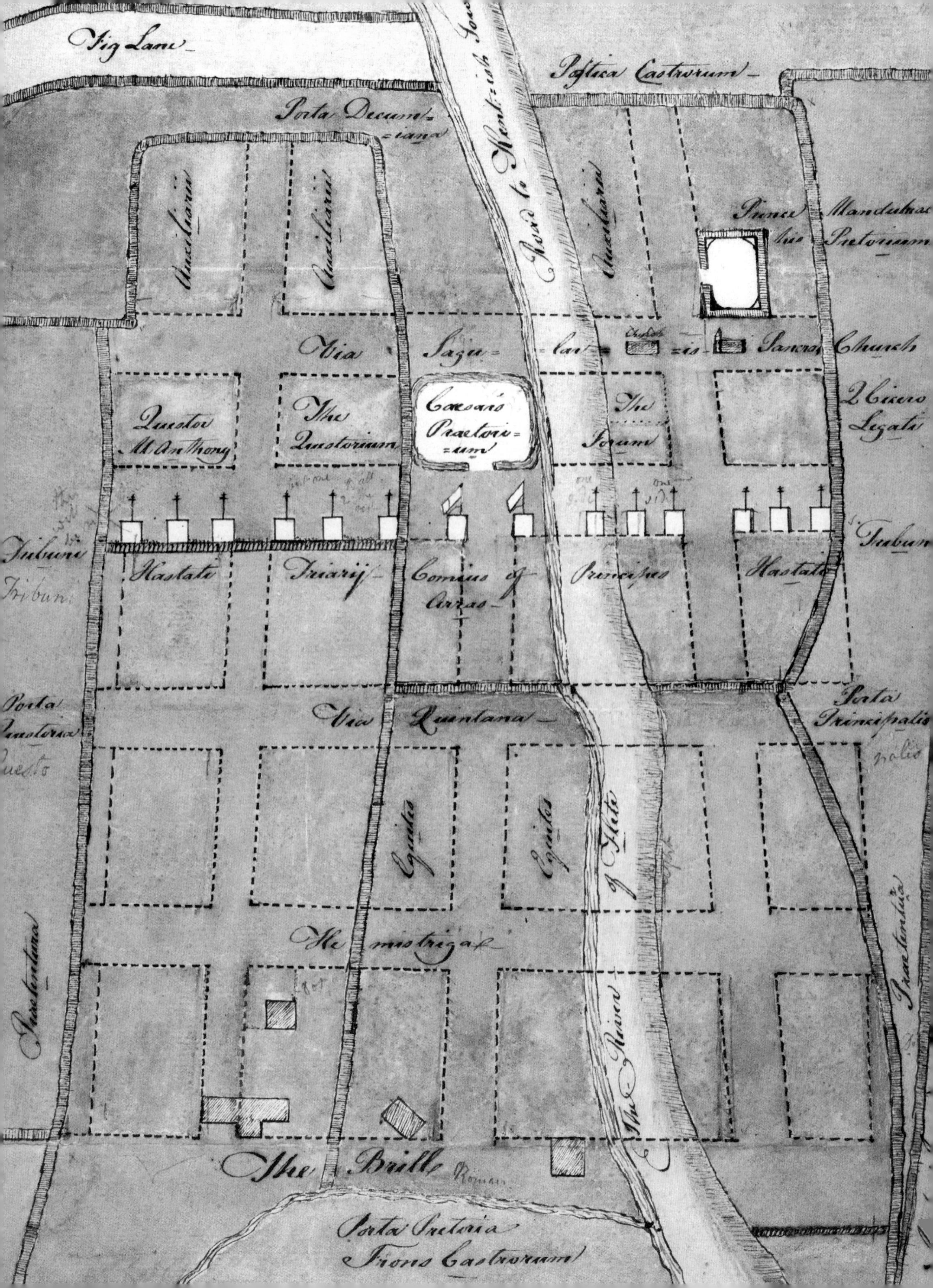

Fig Lane

Posta Decumana

Postica Castrorum

Porta Decuman=
=iana

Road to Mentrath Town

Prince Mandubratii
his Pretorium

Via Sagu-   =lar=   =is   Sacros Church

Caesaris
Praetori=
=um

Questor
M. Anthony

The
Questorium

The
Forum

2 Cicero
Legati

Hastato   Triarij=   Comitis of
Arras   Principes   Hastato

Tribune
Tribuni   Tribun

Porta
Questoria
Questo

Via   Quintana   Porta
Principalis
nalis

Equites   Equites   The Stets   The River

Praetorium   The mistrigas

The Brille   Roman

Porta Pretoria
Frons Castrorum

THE RUDGE CUP

A small bronze bowl found in 1725 down a well in Wiltshire bears one of the few cartographic representations from the Roman province of Britain, an occupation which lasted for over three and a half centuries.

The enamelled bowl, which was unearthed at a former Roman villa at Rudge, is adorned with a series of rectangular grids and crenellations, above which is a brief inscription. These have been interpreted as representing Hadrian's Wall, the great stone barrier which marked the northern edge of the province of Britannia, with the writing referring to the names of five forts at the western end of the Wall; visible here is the name of Camboglans, or Castlesteads fort. Two other metal vessels have subsequently been found (at Amiens, in France, and in Staffordshire) which also list Wall forts (though a slightly different set from the Rudge Cup), suggesting a possible ancient trade in mementoes or souvenirs for soldiers who had served as part of its garrison.

The cup probably dates from the early AD 130s, around a decade after the Emperor Hadrian ordered the wall's construction following a visit to Britain in 122. During it, he decided that his policy of replacing his predecessor Trajan's stance of imperial expansion with one of retrenchment would be extended to Rome's northernmost frontier. In truth, this pull-back had been underway for some time, as the very furthest military outposts at the edge of the Scottish Highlands had been abandoned around AD 86, and forts north of the Solway were evacuated around 103.

The new line ran from Segedunum (Wallsend) on the Tyne in the east to Bowness-on-Solway in the west. At first a stone wall was planned in the eastern section, with a turf rampart providing a barrier in the west of the Wall's 80 Roman mile course. This plan was soon modified to encompass a stone construction along the full length of the wall, with a series of forts built alongside it to house the fortification's garrison. The five listed on the Rudge Cup – Mais (Bowness-on-Solway); Aballava (Burgh-by-Sands); Uxelodunum (Stanwix); Camboglans (Castlesteads), Banna (Birdoswald) – held units of auxiliaries, non-citizen soldiers, of whom some 30–35,000 supplemented the 15,000 men of the three legions based at York, Chester and Caerleon.

The forts along Hadrian's Wall were punctuated by milecastles and turrets, smaller fortifications which held detachments from the main units. The role of this garrison was probably a deterrent and supervisory one, monitoring movements among the tribes to the north of the Wall, controlling passage south in the main area of Roman Britain, exacting tolls from those who passed through its gates and handling any small-scale incursions. Any larger breaches or invasions would be dealt with by pulling troops back from Wall and by the II Augusta Legion based in York.

Hadrian's Wall served its purpose for almost 300 years. From around 140, the border was moved by Emperor Antoninus Pius to a line further north, between the Clyde and the Forth estuaries, where a shorter, turf wall acted as the frontier for some twenty years before Hadrian's Wall was recommissioned. Periodically emperors embarked on punitive expeditions to the north, such as Septimius Severus who came to Britain in 208 to restore order to the frontier, only to die at York. The northern tribes are also recorded as having breached the Wall on several occasions, most notably in 367 when a 'barbarian conspiracy' of Picts, Scots and Irish almost overran the province before being thrown back by Count Theodosius, a special military envoy despatched by the Emperor Valentinian I.

Within half a century, however, Theodosius's work had been undone. Much of the remaining military garrison was withdrawn by Constantine III, a British-based usurper, who took them to the continent in 407 in a failed bid to seize the imperial capital, Rome. In 411, the British expelled the remaining Roman officials and the island was left to its own devices and the mercy of waves of Saxon invaders. Although there is evidence of continued occupation at some forts, including Birdoswald, into the later fifth century, Hadrian's Wall was now redundant and over centuries its towers tumbled and its stones robbed. Only the precious testimony of items such as the Rudge Cup give a sense of the former grandeur of Roman Britannia.

.b.   Ab J

Regnu est anglorum orientalia

Regnu est Serie orientalia

Regnum Northanbron Quod ÷ ma ultimu et magnu

Regnu Int eande oriente et austru

Situs Britannie. et Regna Regulorum.

Regu ÷ orientalium qd ÷ max imu cui pfi rex offa

Regnum Suthsexo qd est austru vale.

Regnum Westsex Quod est orientale

iiii.   ii.   austo

Ab occiden. iii. te

Et si rex offa tanniam lib ut soluf tana reg det iste dus uñe

ALF red SAPI ens

oib; regur b asser. ita i tota bri nare bi tu Alfe monar

MAP OF THE ANGLO-SAXON HEPTARCHY, Matthew Paris

No maps of Britain have survived from the period between the fifth and the eleventh centuries, a crucial time in the nation's development which saw the collapse of the Roman province of Britannia, waves of invasions by Germanic barbarians and, ultimately, the emergence of unitary kingdoms in England and Scotland.

Yet cartographic representations of this formative period do survive from a little later, most notably a map showing the Heptarchy – the seven kingdoms of Anglo-Saxon England – drafted by Matthew Paris, a thirteenth-century Benedictine based at the Abbey of St Albans. Paris was a prolific writer, and the compiler of an important chronicle which covers events from the creation until just before his death in 1259. With access to a major monastic library and important connections at the royal court – St Albans was just a day's ride north of London and Henry III is known to have visited the abbey nine times during Matthew's lifetime – Paris is an important witness to how the English saw the evolution of their nation.

The Heptarchy map – which comes from the *Abbreviatio Chronicorum* – a shortened version of Paris's principal chronicle – is schematic. The seven kingdoms into which Anglo-Saxon England was believed to have been divided (Kent, Sussex, Wessex, Mercia, Northumbria, East Anglia and Essex) are arrayed like petals around a central circle. Although elsewhere he repeats the traditional explanation that all of these had appeared within a single generation, established by the children of Brutus, the legendary Trojan warrior said to have fled from the war with the Greeks, Paris gives them numbers, establishing the primacy of Kent as the earliest Anglo-Saxon kingdom, and Essex as the most recent.

Paris confesses that with the passing of time, it was now difficult to elucidate the dimensions, frontiers and even the ordering of these kingdoms, which dated back to a period (in the seventh and eighth centuries) which was by then some four centuries in the past. His scheme was, indeed an oversimplification. The Angles, Saxons and Jutes who crossed the North Sea to take advantage of the weakness of the native British in the decades following the collapse of Roman authority in 411 did not at first establish neat territorial kingdoms. As the invaders pushed further north and westwards, the war-bands and the authority of local chieftains did coalesce into larger units, which became the kingdoms of the 'Heptarchy'. But even in the eighth century, when King Offa of Mercia (757–96), the revered founder of the Abbey of St Albans, held sway, smaller, less powerful territories survived: notably the lands of the Hwicce and Magonsaete in the West Midlands and Lindsey in the east.

By Paris's time, however, all this had been swept away. The supremacy of Mercia ended when Egbert of Wessex defeated the Mercians in 825 and soon afterwards East Anglia freed itself from their yoke. Viking raiders swept through England from 865, defeating one Anglo-Saxon kingdom after another, until only Alfred the Great's Wessex survived. His children and grandchildren gradually reconquered the Danelaw, the area of the north that the Vikings had occupied, until in 954, York, the invader's last possession, was captured, England was free and – for the first time – ruled by a single kingdom. By then, the Heptarchy was but a distant memory, but one whose image survived strongly enough even 300 years later to be illustrated so eloquently in Matthew Paris's map.

hic abundant leones

mons au reus

Taprobana habet ... nnos bis anno men'a fruges

hic dicitur esse mons ... fug ardens

hic abundant leones

gentes xl ... borrani

Euilach

India inqua sunt gentes xl ...

media

arabia deserca

fluuium

Pison flu mont fumosu

Aracusia

pfidia

arabia

Sina

eudemon

mons sina

flum

chaldea

egiptus superior

mesopotamia

echiniza ... deserta

mare caspium

montes armeni Archa

Pabilo Niua

Siberia

moabite ... mons falgu

amorrei

mons magog

Albanorum

commagona

mons g ...

dimidia tribu manase

ruben gad

phik Plona

colchorum punica

cesara philipi ...

Andoinos

tribu dan

emphorum ge ...

necori del pa dudel

Decisa ... ... cuicas

Issaquar

tribu dan

effraim

mahale

tribus zabu lon

galilea

iericho assos

ebron

mons ...

Selaqua

tribu zabulon

neptalim

mons olimpai

Hierusalem

mons

clinga

Tanai fluuius

nhario cileciae

neptalim

Alexandria

uia minor

cilicia

Scithia

naphida

uia

Libia cirinensis ... caluarsini

melina

attica

Serbia

para abi ec gothia

Tracia

Danubius fluuius

macedonia

penrapo ...

Hunorum gen

achaia ...

Libia etinopu

pannonia alana ...

Zelata

Sardinia

surei

hie argo ... ...

Histria

lauon

Cartago magna

Lagi. rego insa & te offrica ... ... bestis ... portabibus plena mariti ... mica

cinocephales

nero ...

talenza

Britannia londnes

Roma

mauritania

fons ... ... mons

Ispania citerior

insula ... ...

mons achariis

THE ANGLO-SAXON MAP OF THE WORLD

The British Isles sit at the very edge of this early eleventh-century Anglo-Saxon map of the world, as though at the very edge of consciousness. Yet its surprisingly accurate outline of the British coastline marks it out as the product of a vibrant scholarly tradition at a time when England was a cultural power-house, albeit one about to suffer a renewed series of foreign invasions.

As in so many medieval European maps, this one is oriented with east (the direction of Jerusalem) at the top. The British Isles sit at the bottom left-hand edge of the map, whose depiction of fantastic beasts such as the *Gens griphorum* (human-griffon hybrid) seem to indicate it is the product more of monastic fancy than cartographic precision. Yet, though it sits firmly in the tradition of medieval *mappae mundi* (maps of the world), many of which are content to focus on the central position of Jerusalem and to use the maps' surface as a medium for Biblical exegesis, the Anglo-Saxon world map goes further. Its well-drawn outlines (and the presence of accurate Roman provincial boundaries in Asia Minor) suggest that it was drafted by someone who had access to a Roman original, but who incorporated up-to-date information about the coastlines of Britain, Ireland and northern France.

It is possible that the map was composed by someone working in the household of Archbishop Sigeric of Canterbury, who in 990 travelled to Rome to receive his pallium – the archiepiscopal insignia – from the Pope and who on his return wrote an itinerary of his travels. By then, England stood on the brink of a disaster which had been long in the making. The Vikings, who had begun their raids with an attack on the northeastern monastery of Lindisfarne in 793, and had threatened to overwhelm all the Anglo-Saxon kingdoms of England, had been checked by Alfred the Great of Wessex's victory against them at Edington in 878. Alfred's children and grandchildren had gradually retaken land from the Vikings, building a series of *burhs* or fortified towns, beginning with Hertford in 911 and pushing north until they retook the Viking capital of York in 954.

After nearly four decades of peace, the Vikings returned in 991, winning a victory at Maldon in Essex, their bands spreading over the countryside like an unstoppable tide. King Aethelred (whose nickname the 'Unready' means ill-advised rather than unprepared), responded in the worst way imaginable by paying off the raiders with a hefty bribe of £10,000, in part at the suggestion of Archbishop Sigeric. The size of England's ransom served only as an incentive for more raiders to come. By 1012 the payments had ballooned and it took £48,000 to pay off the latest Viking army (though the suggestion that Aethelred might not do so had led to the murder of the latest Archbishop of Canterbury, Aelfeah, who had pleaded with the king not to pay a ransom for him, and for his impudence was pelted by his captors with ox-bones, and killed).

Demoralized and ill-led, the Anglo-Saxon armies rapidly succumbed to the Viking invaders, whose raiding had now become a royal enterprise, directed by King Sweyn of Denmark and his son Cnut. Sweyn briefly ruled England after Aethelred fled, but the definitive Danish conquest happened in 1016 when Cnut overcame and killed Aethelred's son Edmund Ironside. England was now ruled in its entirely by Scandinavian outsiders.

For twenty-five years it was no longer at the edge of the world, but part of a North Sea Empire in which Denmark, England, Norway and parts of Sweden recognized the overlordship of Cnut and his sons. Half a century later, England would suffer a new conquest, at the hands of William of Normandy. The Anglo-Saxon map of the world, therefore, though it was originally composed at a time of comparative peace and tranquil isolation for England, sat the edge of a period of violent confrontation and forced engagement with the rest of Europe.

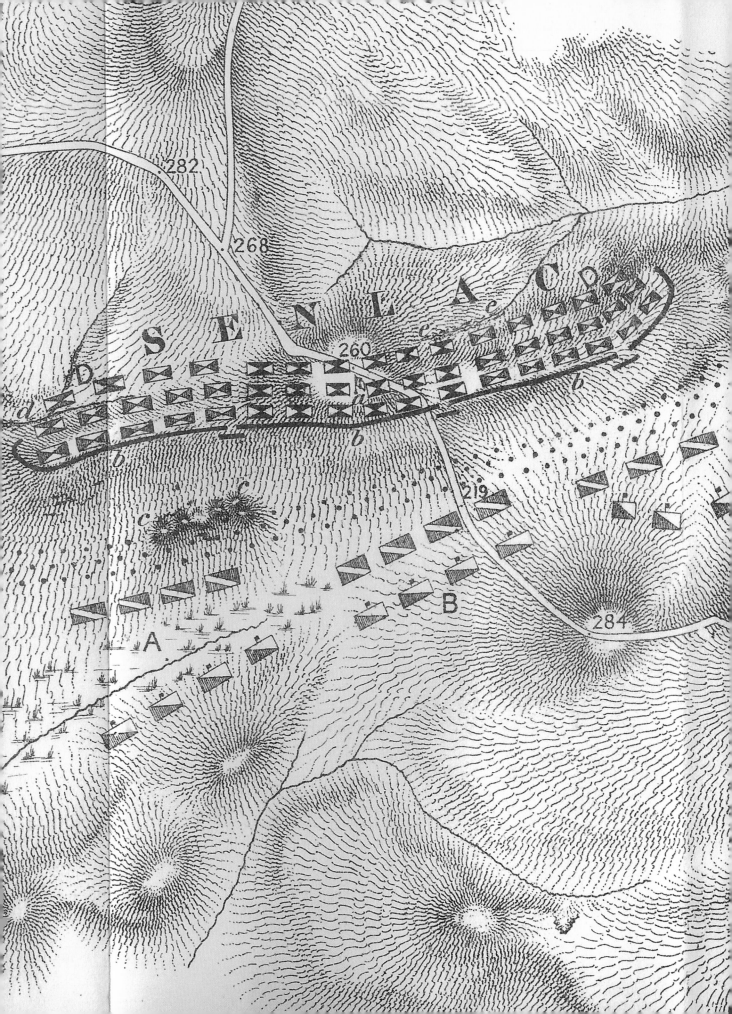

BATTLE OF HASTINGS MAP,
Edward Freeman

The crisply engraved lines in Edward Freeman's map depict one of the greatest turning points in England's history, the Battle of Hastings in 1066. The Norman army of Duke William line up in a great arc below Senlac Hill, where the English warriors of King Harold Godwinson await their charge. To the victor that day the spoils would be the English crown and the chance to shape England's destiny.

Freeman's map was printed in 1869 for the third volume of his *The History of the Norman Conquest of England: Its Causes and Its Results*, some eight centuries after the battle, but there are no contemporary plans of the encounter (though a number of chroniclers who wrote soon after it was fought give a reasonable account of its course).

William's journey to the deployment laid out by Freeman had been a long one. His claim to be England's king was a distant one, as his great-aunt Emma was the wife of King Aethelred the Unready of Wessex. More importantly William, who had ruled Normandy since 1035, maintained that he had been promised the English throne both by Edward the Confessor (who, before he became king in 1042, had spent time as an exile in Normandy as a boy) and by Harold Godwinson, who had been an enforced guest of William after being shipwrecked in northern France, probably in 1064.

When Edward died in January 1066 and the English nobility selected Harold to succeed him, William was furious and began to plot an invasion. More immediately pressing from Harold's point of view, however, was another invading army which had landed in northern England. Led by Harald Hardrada of Norway, who believed that he had inherited the rights of the Danish dynasty which had ruled England until 1042, it defeated the local English militia under Earls Edwin and Morcar and occupied York. Harold led the English fyrd, the national levied army, on a forced march northwards and caught Hardrada by surprise, defeating and killing him at Stamford Bridge, near York.

By now Harold had learnt that William had assembled an invasion fleet, slipped across the Channel and landed on the south coast near Pevensey with an army of around 8,000 men.

He was forced to turn back south and intercept the Norman duke before he could reach London. A long march brought Harold's host to Sussex and then, after resting the night, he arrayed his army, which was about the same size as William's, along a ridge known as Senlac. In the fashion typical of Anglo-Saxon (or Viking) hosts they arrayed themselves into a shield-wall, the overlapping of their shields creating an almost impenetrable barrier, which bristled with spears.

William ordered volley after volley from his archers and then infantry charges which failed to make any dent in the shield-wall. His men's morale almost crumbled when he was unhorsed and one wing of Bretons fled, to be followed by a detachment of Anglo-Saxons who chased them down off the hill. Yet this proved the Normans' salvation as the Bretons rallied and cut their pursuers to pieces. Further feigned flights tempted more of the defenders from the safety of the shield-wall. Still, the Anglo-Saxon line may have held, but, as time was running short for William to force a victory, disaster struck. A stray arrow hit Harold in the eye, fatally wounding him. The morale of the Anglo-Saxon fyrd collapsed and – with Harold's brothers also dead – they fled, and hundreds of them were butchered during the retreat. After that, resistance was sporadic and William was left in control of the battlefield. Although the Witan, the English royal council, tried to offer the throne to Edgar Aetheling, another Anglo-Saxon prince, he commanded little support and no army, and when William approached London, he too submitted.

William was crowned at Westminster Abbey on Christmas Day 1066 in a church that had been inaugurated almost exactly a year earlier while Edward the Confessor was king. The event was nearly a disaster as Norman soldiers attending the coronation, hearing shouts of acclamation for the new king, mistook them for an uprising and set fire to houses around the Abbey, causing the church to fill with smoke and many of those attending fled in panic. Despite the inauspicious start, William consolidated his control over England, which was seriously threatened only by a major rebellion in the north of England in 1069–70. His descendants would still be ruling over England 950 years after the Battle of Hastings.

see more on next page >

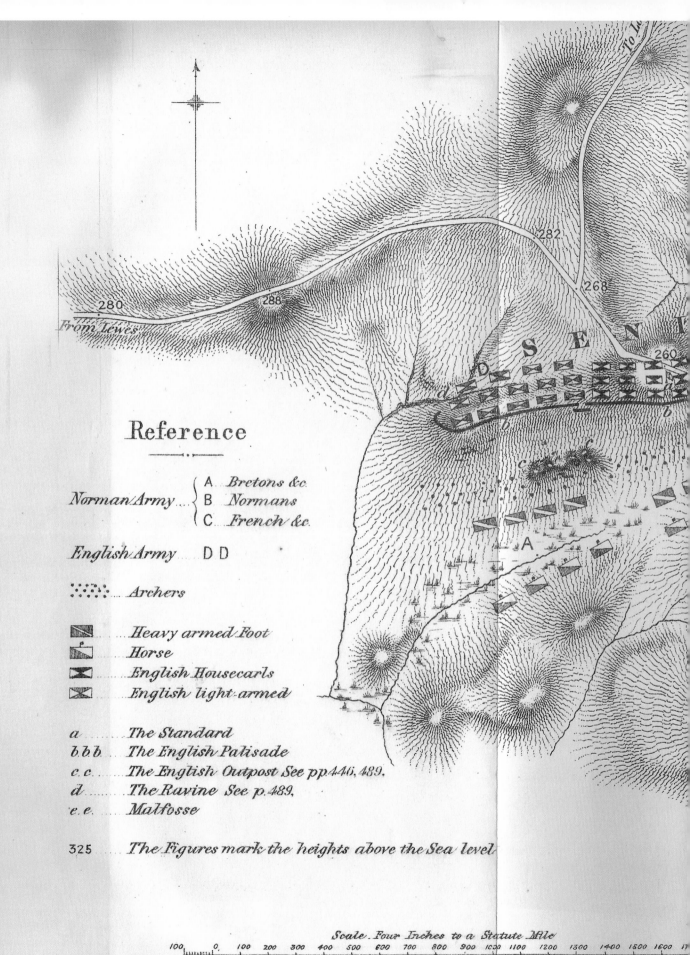

Reference

Norman Army — { A Bretons &c.
                B Normans
                C French &c.

English Army — D D

⋰⋱⋰ Archers

Heavy armed Foot
Horse
English Housecarls
English light-armed

a .... The Standard
b.b.b. The English Palisade
c.c.   The English Outpost See pp.446, 489.
d .... The Ravine See p.489.
e.e.   Malfosse

325   The Figures mark the heights above the Sea level

Scale. Four Inches to a Statute Mile
100 0 100 200 300 400 500 600 700 800 900 1000 1100 1200 1300 1400 1500 1600 1700

Zincographed at the Ordnance Survey Office Southampton Major General Sir Henry James R.
1874

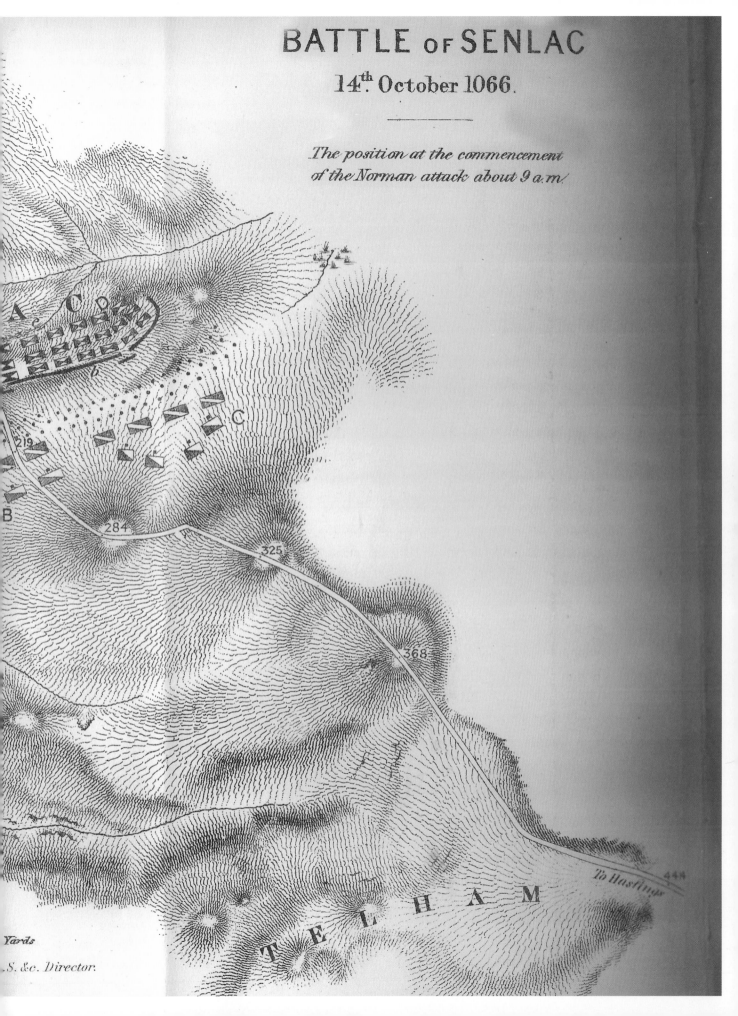

BATTLE of SENLAC
14th October 1066.

*The position at the commencement
of the Norman attack about 9 a.m.*

A
C
D

219

B

284

325

368

To Hastings

444

TELHAM

Yards

S. &c. Director.

TOPOGRAPHIA HIBERNIAE,
Gerald of Wales

Among the earliest visual representations of Ireland, Gerald of Wales's *c.*1188 map is highly schematic, providing few details of the geography of the island. Ireland is merely an amorphous pinch-waisted blob ('Hybernia') floating below 'Britannia' and next to the incongruously large Orkneys ('Orcadia'). Yet the accompanying text, setting out a natural history and account of the peoples and customs of Ireland, betrayed attitudes which were to bedevil the relationship between English and Irish for centuries.

Gerald was a member of the aristocratic Norman Fitzgerald family, who in common with many ambitious aristocrats had carved themselves out a demesne in the English–Welsh border marches. He first went to Ireland to visit members of his family who had settled there after the invasion by a party of Norman knights in 1170, and was then sent to accompany Prince John, Henry II's younger son, in 1185. Gerald seems to have visited Waterford, Cork, Dublin, Wicklow, Meath and Kildare and, on his return, concocted a racy mix of folklore, confabulation, hearsay and personal observation to create his *Topographia Hiberniae*. Interleaved with traditional tales of the High Kings of Ireland (which are not wholly divorced from reality) are stories of a bearded woman with a mane on her back, a hybrid creature who was half-ox and half-man, and a fish that had been found with three golden teeth. About the Irish he is scathing, portraying them as degenerate savages who exhibit 'abominable treachery' in their dealings with others.

If not exactly treachery, it was certainly an act of betrayal that first brought the English to Ireland. In the 1150s, Dermot MacMurrough, the King of Leinster, was locked in a long struggle with Rory O'Connor, King of Connaught, and Tiernán O'Rourke of Breifne, whose wife, Dervorgilla, Dermot rather unchivalrously kidnapped in 1152. Already, the English crown had become interested in exerting control over Ireland, and in 1155 Henry II induced Pope Adrian IV – who was, not coincidentally, from England (and the only English-born Pope) – to grant him a Papal Bull, *Laudabiliter*, which gave the sanction of the Church to the conquest of Ireland by Henry any time he chose to exercise it.

Henry bided his time, and even when Dermot found himself exiled from Ireland after Rory O'Connor captured Dublin in 1166 and had himself crowned High King, he did not intervene. Instead he allowed the exiled Leinster king to recruit a mercenary force among the Anglo-Norman nobility. With the inducement of a promised marriage to Dermot's daughter Aiofe, and the prospect of succeeding to this throne in due course, Richard fitz Gilbert de Clare, Earl of Pembroke (known as 'Strongbow') led a small party of knights to Ireland in 1170. They succeeded beyond their wildest expectations, and so concerned was Henry that an independent Norman kingdom might arise across the Irish Sea that he intervened in person, landing in Waterford in October 1171.

Having received tribute from most of the Irish kings, and placed a series of garrisons, including at Cork, Limerick and Dublin, Henry returned to England six months later. By the Treaty of Windsor in 1175 he recognized Rory as High King of Ireland, but in strict subordination to Henry, who took the title Lord of Ireland. The English tried to undermine his position even further, when Henry petitioned Pope Lucius III to have his son, John, crowned as Irish King. Although this failed – Lucius was an Italian Cistercian with fewer reasons to oblige the English crown – John was sent in 1185 to visit the land over which he would not now be king. Gerald of Wales went with him, and his attitude to the Irish was perhaps pervasive amongst John's entourage. At Waterford, some of the English courtiers pulled the beard of the Irish chieftains who came to greet the English prince, and after that none arrived at other towns to offer him fealty.

Step by step the English hold on Ireland tightened. Royal officials were appointed, such as a Treasurer of Ireland in 1217, and English-style counties established (Kildare in 1297 and Carlow in 1306, for example). Irish resistance was fitful, hampered by perennial in-fighting between the Irish kingdoms, and appeals to foreign lords to take the Irish crown such as Haakon IV of Norway in 1263 and Edward, brother of Robert Bruce (who did launch an expedition in 1315) met with limited success. Only the mellowing of the Anglo-Norman families as they partly blended with their Gaelic neighbours and became distant from the priorities of the Westminster court slightly softened the harshness of the English regime in Ireland.

When the Statue of Kilkenny was passed in 1366 it was an attempt to reverse this trend. Intermarriage between English and Irish was strictly forbidden and the English colonists in Ireland were to be governed by English common law. The Irish were reduced to the status of a dangerous 'other' to be kept strictly at bay. It was an attitude which had its roots in Gerald of Wales and his tales of fantastic beasts and barbarous customs.

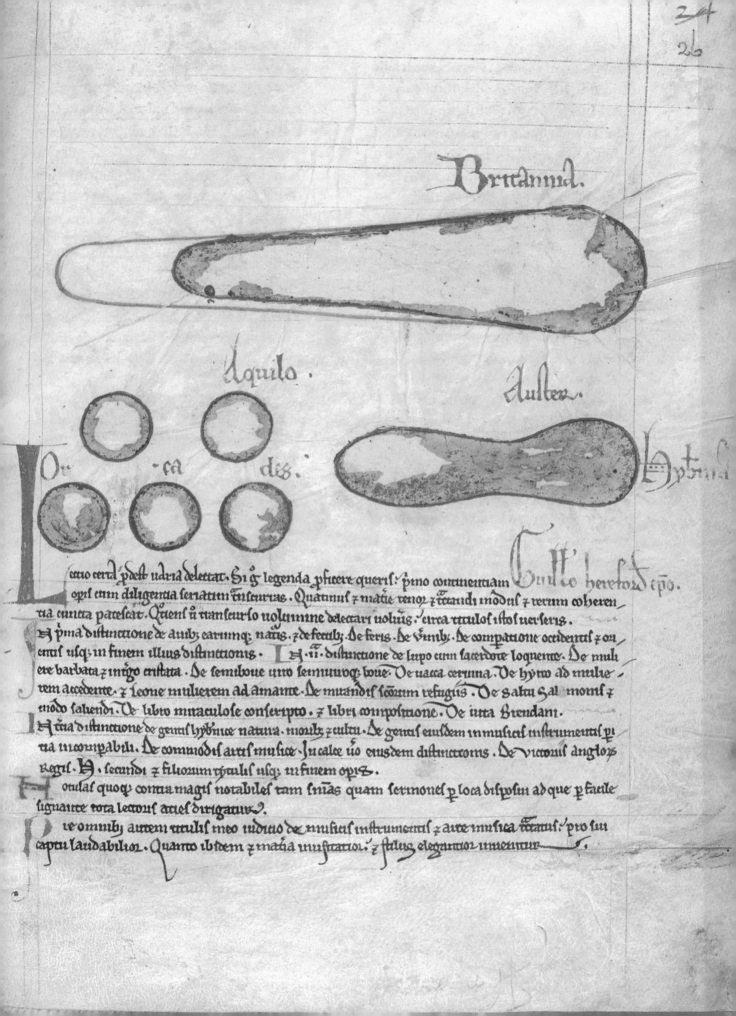

Britannia.

Aquilo.

Auster.

L Or ca des. Hybnia

Guitto hereforo epo.

L ecuo certa p̄dest udra delectat. Si q̄ legenda p̄sicere queris: p̄mo continentiam
opis cum diligentia seriatim trīscurras. Quatinus ⁊ matie tenor ⁊ tractandi modus ⁊ rerum coheren
tia cuncta patescat. Quiens ū transcurso uolumine delectari noluis: circa titulos istos uerseris.
H p̄ma distinctione de auib; earūq; natis. ⁊ de feculis. De feris. De serpē. De comparatione occidentis ⁊ on
entis usq; in finem illius distinctionis. H .ij. distinctione de lupo cum sacerdote loquente. De muli
ere barbata ⁊ iuigo cristata. De semiboue uiro semiuiroq; boue. De uacca ceruina. De hyrco ad mulie
rem accedente. ⁊ leone mulierem ad amante. De mirandis scōrum refugiis. De saltu Sal montis ⁊
modo saliendi. De libro miraculose conscripto. ⁊ libri compositione. De uitta Brendani.
H tcia distinctione de gentis hybnice natura. moribz ⁊ cultu. De gentis eiusdem in musicis instrumentis p̄
tia incompabili. De commodis artis musice. In calce uo eiusdem distinctronis. De uictoriis anglox
regis. H secundi ⁊ filiorum titulis usq; in finem opis.
H otulas quoq; contra magis notabiles tam sn̄as quam sermones p loca disposui ad que p facile
signante tota lectoris acies dirigatur.
P re omnibz autem titulis meo iudicio de musicis instrumentis ⁊ arte musica tractatus: pro sui
captu laudabilior. Quanto ibidem ⁊ matia inustracior ⁊ stilus elegantior inuenitur

MAP OF BRITAIN, Matthew Paris

Perhaps the earliest reasonably accurate cartographic representation of Britain, the map created by the St Albans monk Matthew Paris around 1250 also marked a time of transition between the political units that made up the island, as a newly assertive England attempted to absorb both Wales and Scotland within a single kingdom.

Paris included the map in the *Abbreviatio Chronicorum Angliae*, a shortened version of his major chronicle of English history. It includes over 250 place-names, the major towns such as London, being indicated by stylized fortifications (and occasionally churches). Important rivers are also shown, as well as the two Roman walls: Hadrian's Wall running westward from the mouth of the Tyne and the more northerly Antonine Wall bridging the gap between the Firths of Clyde and Forth.

Although generally accurate (and certainly so in comparison to most medieval maps), the geography of Britain is distorted. Matthew could have had a Roman map as a model, which may help explain the relative paucity of information shown about Scotland and its unnaturally small size. The heart of the map is also based on an itinerary from Newcastle to Dover via London, leading cities along this spine being placed on a north–south line, displacing Kent, so that it is shown to the south of London, while artificially enlarging East Anglia. Paris was aware of this shortcoming and wrote on another version of this map that, had the page allowed it, the whole island of Britain should have been longer.

The whole map, however, almost self-consciously presents Britain as a whole, despite its division into the political units of England, Scotland and Wales. The thirteenth century was a time when a notion of 'Britannia' was becoming firmly embedded in the minds of historians and polemicists (or, at least, of English ones). Although England was at peace with Scotland and Wales during the reign of Henry III (1216–72), those good relations were brutally shattered during that of his son, Edward I (1272–1307).

Edward, nicknamed 'Longshanks' for his remarkable stature, was a strong-minded and determined ruler. He was not afraid to challenge the church – the Statute of Mortmain (1279) outlawed the granting of lands to monasteries – or the barons –

the Statute of Gloucester (1278) curtailed the liberties enjoyed by landholders, by forcing them to justify by what right or warrant ('*quo warranto*') they had judicial or financial rights within their holdings.

Edward was equally forceful in dealing with his neighbours. During the chaos of the civil war which had almost torn the realm apart in the 1260s, the Welsh ruler, Llywelyn ap Gruffydd of Gwynedd, had taken advantage of England's weakness by securing his recognition as Prince of Wales. Edward was determined to revoke this privilege. When Llywelyn refused to pay him homage in 1275, he launched an invasion two years later which severely reduced Gwynedd's territory and, after Llywelyn rebelled in 1282, led to the complete annexation of Wales. In 1284, the Statute of Rhuddlan declared that the whole principality was a possession of the English Crown.

Edward's war against Scotland was more protracted and ultimately less successful. The opportunity for English intervention arose because of the death in quick succession of the King of Scots, Alexander III, in 1286, and then his granddaughter and heir, the seven-year old Margaret of Norway, in 1290, a double blow which left over a dozen claimants to the throne. Fearing a civil war between John Balliol and Robert Bruce, the principal contenders, the Scottish nobility invited Edward I north to arbitrate. Although he did so, and ruled that Balliol should have the throne, it soon became clear that his real motive was to reduce Scotland to the status of an English vassal. Balliol tried to resist and negotiated an alliance with the French in 1295 (beginning the long history of the 'Auld Alliance' between the two countries) but Edward grew impatient and in 1296 launched an invasion.

This was the first of nine English armies that marched north between then and Edward's death in 1307. By then Robert Bruce had emerged as the main champion of Scottish independence, but his ultimate victory at Bannockburn in 1314 and the driving of the English from Scotland would have seemed an unlikely prospect at the time. The alternative vision presented in Matthew Paris's map, of a Britannia united, must have appeared a much more plausible version of the future.

Le chastel de Doure lentree e la clef
de la riche isle de engleter. e au

l'abbeie seint
augustin

Cantebire. chef de iglises de engletere

ken
t

lettre de Ajedeweie

Roueecestre ki est eueesches

MVSEVM
BRITAN
NICVM

La cite de lund' ki est chef denglere.
Brutus ki prime enhabita engleterre
la funda. E apela troie la nuuele.

la tun
la grant Riue de tamise
la .E. punt
entre
la iglise sei pol

lamheth
Westm'
seint mara

Pois

Seit entin

Seint Richen

Arraz

Musterol

Cateis

Nredame de
Boeloine

Wifan't port
de mer cunt
Doure

ITINERARY MAP, Matthew Paris

In his mid-thirteenth century chronicle of world history, the St Albans monk Matthew Paris included an itinerary map showing the routes from England to the Holy Land. It is both a reflection of the religious sensibility of the age and of very real political problems which dogged the long reign of his royal master, Henry III (1216–72).

At the beginning of his *Chronica Majora*, Paris included seven pages of maps showing the journey pilgrims might take from London to Jerusalem. Their strip-like composition is very much in the tradition of late Roman itinerary maps, but the final destination, the sacred city of Jerusalem, indicates that the map had a spiritual purpose. Paris had never been to the Palestine and – given the hardships of travelling there, involving a journey of many months and the increasing pressure that the Christian crusader states in the Levant were coming under from local Muslim rulers – very few of his contemporaries had, either.

Yet his map allowed an imaginary pilgrimage, tracing the route all the way from London – the image of which at the foot of the page, with the spire of St Paul's Cathedral conspicuous in the centre and the Tower of London (the '*Tur*') and London Bridge (the '*Punt*') among the prominent landmarks, is one of the earliest portrayals that survive of the English capital. The map's detail is most intense in its earlier segments, until it reaches Apulia in southern Italy, a feature which has led its composition to be linked with the crusading expedition of Richard, Earl of Cornwall, the brother of Henry III. On his return from the Holy Land, he disembarked in Trapani in Sicily in 1241 and returned overland, a journey he is thought to have described to Matthew Paris.

Richard was invited to renew his acquaintance with southern Italy in 1252, when Pope Innocent IV offered him the crown of Sicily as a means of disposing of his hated Hohenstaufen rivals who controlled the island but were weakened after the death of their great champion Holy Roman Emperor Frederick II two years previously. The whole exercise seemed so quixotic that Richard turned him down, saying the Pope might just as well have invited him to take flight and capture the Moon. Henry III, though, became committed to the enterprise, this time in the name of his younger son, Edmund, and the costs of preparing a campaign in Sicily, proved crippling. By 1258, when it was fairly clear that the 'Sicilian Venture' would never come to fruition, Henry had amassed debts of £100,000 and, perhaps worse, faced the threat of excommunication for failing to adhere to his vows to the Pope. Already unpopular for the high taxation which he had previously imposed and for his generally autocratic rule, Henry faced a storm of protest when he summoned a parliament at Oxford in June 1258 and asked it for a levy to resolve his funding crisis. The assembled barons countered by presenting him with the 'Provisions of Oxford', a document which would have neutered royal power by placing real power in an assembly of fifteen of their own number and providing for parliament to meet three times a year as a check on Henry's authority.

Both Henry and the baronial leader, Simon de Montfort refused to compromise and civil war broke out. Although the king suffered an early military setback at Lewes in 1264, he recovered thanks to a crushing victory won by his son Edward at Evesham the following year, when de Montfort was killed. By then, however, Henry had been forced to sign the Treaty of Paris (1259) conceding most of northern France to the French king, Louis IX. Meanwhile, the parliament which de Montfort had called in London in 1265 included the lesser orders, knights and burghers from the main towns, widening parliamentary representation in a way that subsequent monarchs were never quite able to undo.

Even before Matthew Paris completed his map, the road to Jerusalem was blocked once more, as the city had fallen to the Khwarezm Shahs, central Asian mercenaries, in 1244, never again to be recaptured by Christian crusaders. And the journey to Sicily, which the Pope had so invitingly extended to Henry III's brother, had ended with the dissipation of English power in France and a parliamentary affront to royal authority which would have very serious long-term consequences for England's constitutional development.

THE HEREFORD MAPPA MUNDI

The late thirteenth-century *mappa mundi* ('map of the world') at Hereford Cathedral depicts Britain as a compressed oblong at the corner of the known world, a landscape which is instead dominated by the sacred city of Jerusalem, placed boldly at the centre of the map, mirroring its vital role for Christian devotion and religious sensibility.

As knowledge of Roman cartography withered away in medieval Europe, maps became more than ever a way of presenting a Christian vision of the world in geographic form, drawing deeply on a combination of classical scholarship, the Bible itself and the works of early Christian writers. The main concerns that emerged in Late Antique and Early Christian Europe were the portrayal of the topography of the Holy Land, where Christ's story had unfolded, and a broader attempt to frame maps of the world in a Christian context. These latter came to be referred to as mappae mundi, from the Latin mappa (or 'cloth') and mundus (meaning 'world') and around 1,100 of them have survived.

The Hereford Mappa Mundi is the summit of a tradition which began with much simpler maps, known as T–O which portray the world as a flat circular disk. In these, the world is surrounded by an Ocean, with the body of the T dividing the land into the continents of Asia (generally the largest, taking up the whole top half of the map). The Hereford map was compiled around 1290, and that the map's intention is the presentation of the Christian history is made clear by the figure of Christ at the top, sitting in judgement over virtuous souls, who are led upwards to heaven by the angels, and a cohort of the damned, who are dragged by demons to the burning mouth of Hell.

Christian topography is further privileged by the setting of the Earthly Paradise – from which Adam and Eve were expelled - at the top (far east) of the map and the disproportionately large drawing of Jerusalem at its very centre.

Britain seems a marginal place, not even graced by any of the fantastic beasts and mythical figures which fill blanks in the map at the fringes of Asia and Africa (many of them drawn from the map's classical antecedents such as the works of the Roman naturalist Pliny the Elder and the fifth-century historian Orosius). Only about forty place-names are given in England, Scotland and Wales and a scattering of towns, with the multi-turreted London being the strongest visual presence. Even the town of Hereford, where the map came to be stored, has been added as an afterthought (together with the line of the River Wye), perhaps because it is thought that the map was originally created at Lincoln.

There is no real attempt at geographical accuracy; Wales is unnaturally elongated and Scotland is represented as an island, sundered from northern England by a deep channel. Yet the author, who has been various identified as Richard of 'Haldingham or Lafford' or Richard de Bello, places next to Britain the figure of the Roman Emperor Augustus despatching four surveyors to compile a map of the known world. It is a way of resolving the paradox that, though classical history and religious tradition had no central role for Britain, for scholars working there – and for their noble and ecclesiastical patrons – a way had to be found invest their country with a privileged position in the terrestrial order.

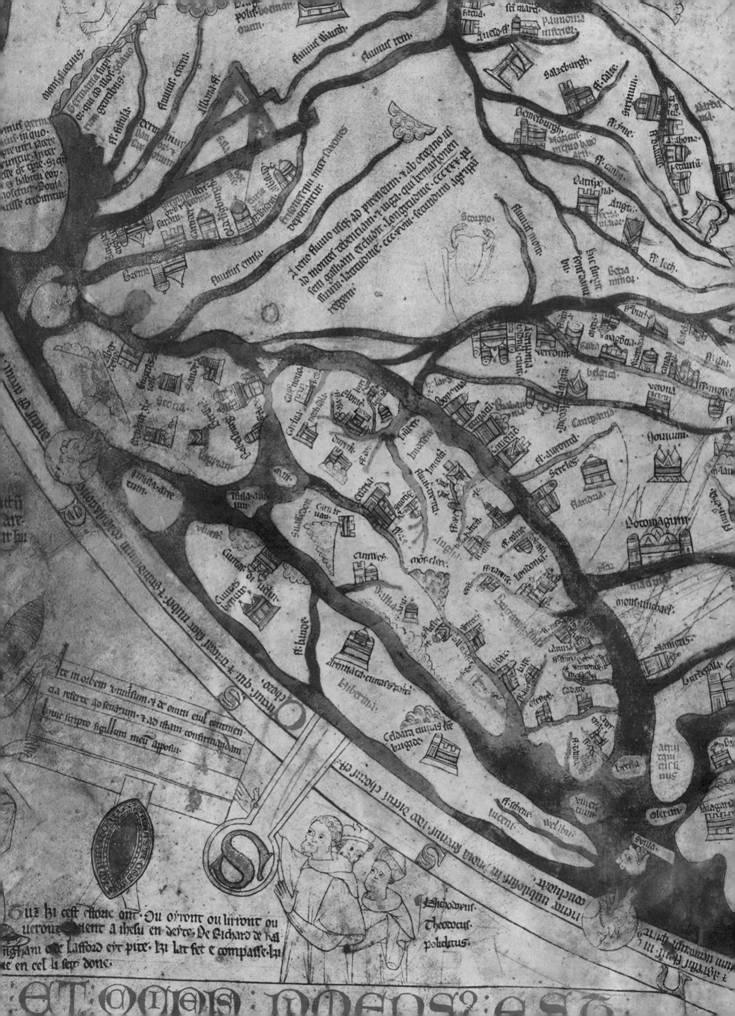

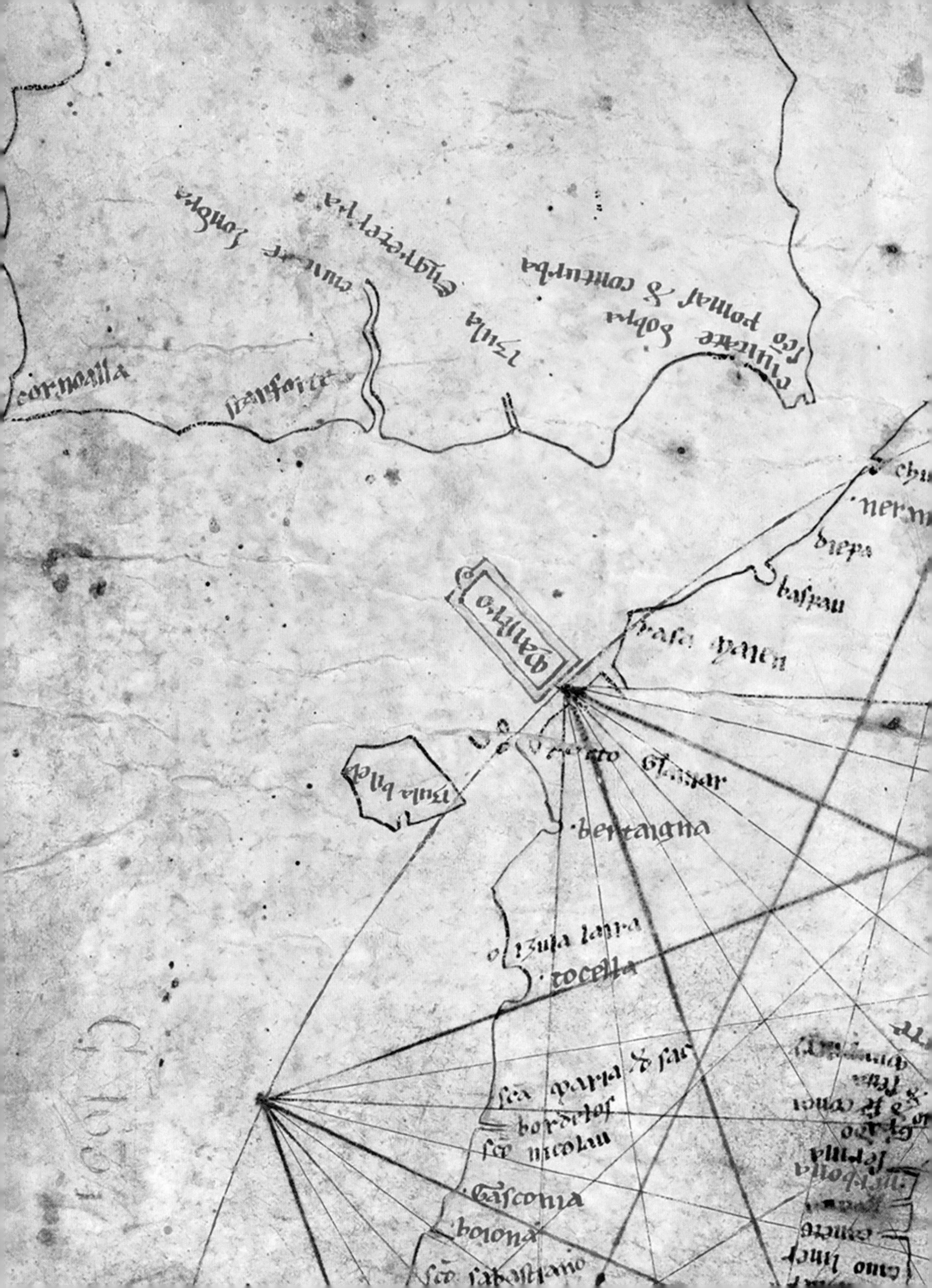

cornoalla

transfota

ysula Engleterra

civitate zobra
leo pollas & conturba

chu
nerm
diepa
balsau

rrafa palen

FANTRO

Serrado clossar

ysula blea

bertaigna

ysula lanra
rocella

sca maria d sar
bordelof
sco nicolau

Gasconia
bolona
sco sabastiano

CARTE PISANE

Britain barely features in one of the earliest European maps intended for practical use. The outline of the British coastline at the top left of the Carte Pisane, a late thirteenth-century portolan (or mariner's chart), is schematic in the extreme and includes only six place-names along the south coast. Yet it is a sign that, no matter how peripheral, Britain's insular status did not equate to isolation from political, diplomatic and cultural currents that were sweeping through Europe.

Portolan charts differ significantly from the medieval *mappae mundi* which preceded them; their purpose was not religious exegesis or to enable pilgrimages to the Holy Land, but to provide a practical guide for mariners, particularly those travelling through the Mediterranean. The network of lines which permeate the map's surface (at times making it hard to make out the shape of the land masses) are intended to indicate the shortest sailing route between ports, while the setting of the place names on land at an angle perpendicular to the coast show that this is a sailor's eye view of the continent, exhibiting little interest in the interior.

The Carte Pisane – named for the Italian port city of Pisa where it was found – dates from around 1290. The density of place-names in the Mediterranean, the home ports of the Pisan, Genoese and Venetian seamen who were carving out maritime empires for their mother cities, is in stark contrast to England, where of the half-dozen locations named, the most prominent, Civitate Londra (or London), is misplaced on the south coast, roughly where Southampton should be. The map also lacks the array of manticores, wyverns and other fantastical beasts which adorned its medieval predecessors and the only concession to religious sensibility is a small cross next to the town of Acre in Palestine (not shown on the extract illustrated here).

Acre was the last crusader fortress to fall back into Muslim hands, in 1291, at around the time the Carte Pisane was drawn. The crusading movement had been one in which England most definitely formed part of the European mainstream – mirroring the general alternation of extreme enthusiasm and deep apathy that deprived the crusader states of the consistent support which might have sustained them. Indeed, when Acre fell, an Englishman, Otto of Grandson, was leading a small continent of knights there as a substitute for his sovereign, Edward I, who had sworn several times to take up the cross, but had found his wars against Wales and Scotland too time-consuming to carry out his vow.

Edward had, though, as Prince of Wales, already led a small-scale crusading expedition, travelling to Tunis with 1,000 English knights and men-at-arms to join Louis IX of France in

his attempt to seize the strategic North African port. The French king had most unfortunately died shortly before Edward got there, and so he proceeded to Palestine with the tiny English army and a few French noblemen still eager for glory. He arrived in the Holy Land in 1271, but soon found his company too small to make the slightest difference in the face of the Mamluk forces of Baibars, sultan of Egypt, and after a few months, having made the face-saving gesture of reinforcing the walls of Acre with a tower, he boarded ship back for England.

Further back still, the English crusading spirit had really sparked to life when Richard I led an army of around 8,000 to join the pan-European crusade to liberate Jerusalem (which had fallen into the hands of Saladin, the Ayyubid sultan of Egypt, in 1187). The crusade had already been preached in England, and a hefty series of taxes levied (which became known as the 'Saladin tithe'). However, the expedition was put on hold because the poisonous state of relations between Richard, and his father, Henry II, which meant that it was not considered opportune to send so many fighting men to Palestine. Only after Richard and his ally, Philip Augustus of France, ambushed the old king near Chinon, and an exhausted and bitter Henry died soon thereafter, was the crusade able to proceed.

Although the German Emperor Frederick Barbarossa drowned while crossing a river in Anatolia, and the large German contingent then returned home, Philip Augustus and Richard reached the Holy Land safely. Richard put an end to Saladin's winning streak at Arsuf in September 1190, when a charge by the Knights Hospitaller blunted the Muslim advance and allowed the rest of the crusaders to counter-attack. Although the peace deal he subsequently brokered with Saladin did not restore Jerusalem to Christian control (merely allowing safe passage to unarmed pilgrims), it did ensure Richard's place in English crusading mythology as the saviour of the holy places (and his nickname as 'the Lionheart'). The fact that he barely spent any of his reign in England (being taken prisoner on his way back by Leopold V of Austria, who believed Richard had been complicit in the murder of his cousin, and whose release cost his subjects 150,000 marks in ransom money) was conveniently forgotten.

England was entangled with Europe in other ways, too. Ever since the Norman conquest of 1066, English kings had held lands in France (initially as part of the Norman patrimony and then with a significant expansion in the southwest after the marriage of Eleanor of Aquitaine to Henry II in 1152). Although most were lost by King John in 1215, a subsequent recovery – in ambition if not in territory – would lead to Edward III claiming to be king of France, and the outbreak of the Hundred Years' War in France in 1337. If not an effective guide to England, the Carte Pisane might at least act as one for Englishmen travelling to the continent.

see more on next page >

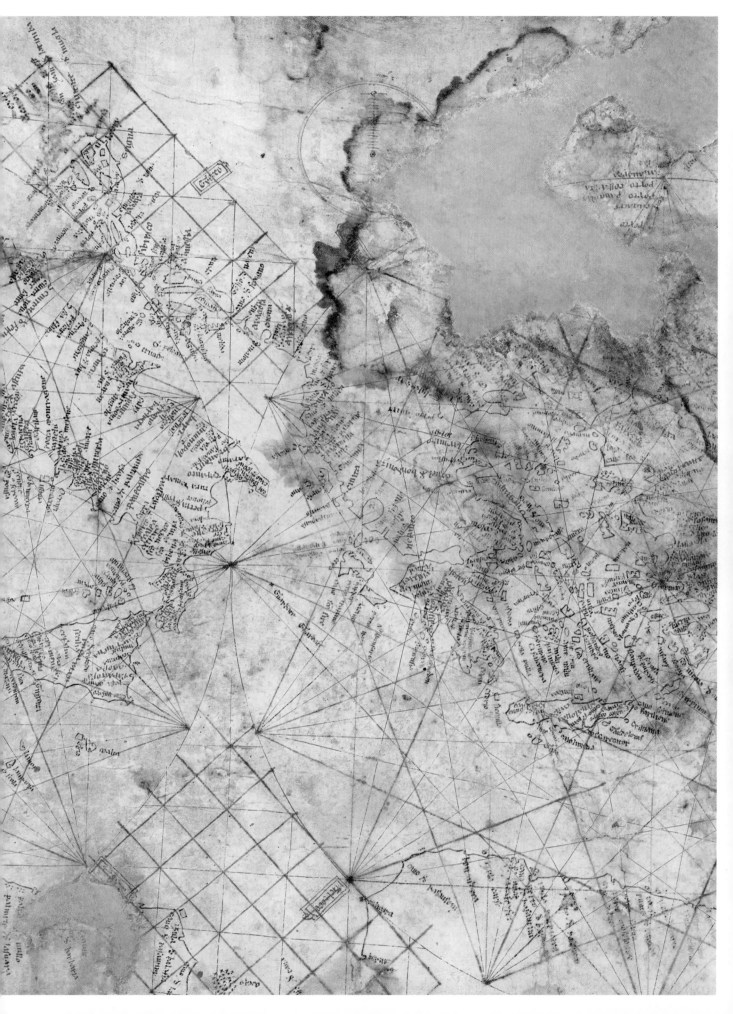

THE GOUGH MAP
OF BRITAIN

Its once vibrant colours faded with age to a dull wash, this mid-fourteenth century map of Britain is a unique document. The Gough Map – named for Richard Gough, who donated it to the Bodleian Library at Oxford University in 1809, rather than for its unknown maker – is the first proper geographic representation devoted to Britain alone. It marks a change in the nation's perception of itself at a time it was undergoing profound demographic change after the catastrophe of the Black Death.

The Gough Map is something of an enigma. Its maker, patron and even its purpose are unknown, but at 0.6 m by 1.2 m (2 ft by 4 ft), made up of two pieces of hide sewn together, and containing over 600 place names and a network of routes connecting principal settlements, its compilation must have entailed considerable effort. The portrayal of the city walls of Coventry, which were only begun in 1355, suggest an earliest date of the 1360s for the map, although equally the cartographer's particular interest in North Wales had led to suggestions that part at least of the information for the map was gleaned during Edward I's campaigns into Wales between 1277 and 1284.

The map is a departure from previous medieval *mappae mundi*, which privileged religious concerns over any need for geographic accuracy. Even though the outlines of Scotland and Wales are grossly oversimplified (and that of Cornwall partly so), the wealth of geographic information (including rivers) is vastly greater than that of any of its predecessors. The red lines which spread throughout the map focus on the five principal routes out of London and may owe their heritage to a tradition of Roman itineraries, which were intended to guide travellers along a route rather than to show accurately all possible means of travelling around a region. It does, for example, miss out many major roads which were known to have existed (such as the Fosse Way).

An annotation in the sea near Devon which reads *hic Brutus applicuit cum Troianis* ('here Brutus landed with the Trojans') refers to the patriotic legend that Britain owed its origin to Brutus, a descendent of Aeneas, a Trojan hero who had fled the sack of his city by the Greeks. Yet it was also in the West of England that another traveller arrived in the summer of 1348, whose progress would transform the country, but with devastating effect.

In early June 1348 a ship, probably carrying wine from Bordeaux, docked at Melcombe Regis in Dorset. It bore a deadly additional cargo, for at least one of its crew was infected with the plague. The disease, to be known to posterity as the Black Death for the swellings and black blotches which affected its sufferers, was carried by fleas which contracted it from rats and then passed it on to humans by biting them. It had spread from Central Asia to Constantinople by 1347 and then took advantage of the Mediterranean trade routes to pass through southern Europe and move rapidly northwards.

Soon people began to die. There was no cure; despite the invocation of all manner of remedies, such as onions, treacle or even posies of cinnamon and roses, nothing helped the patients, though prayer (another suggestion) did at least make them feel better and fumigation (also advised) may have kept away the fleas that were the plague's principal vector. There was panic as the Black Death moved steadily through Britain, reaching London by January 1349, Abergavenny in Wales by May and Scotland by early 1350. The disease struck rich and poor indiscriminately and even the clergy were not immune, with just under half the priests in the diocese of Bath and Wells falling victim.

By the time the disease had run its course, at least a third, and possibly as many as a half, of the population had died (or around two to three million people in England and Wales). Agricultural land fell vacant as there were not enough tenants to till the land, and, with labour scarce, the peasantry began to demand higher wages and better conditions. So concerned did the government become at this trend, that in 1351 a Statute of Labourers was passed forbidding the paying of wages at levels any higher than they had been before the plague. In turn, the peasantry resisted this restriction on their new freedom of action, contributing to further repression and violent outbursts such as the Peasants' Revolt in England in 1381.

England was a more volatile place as the bonds of feudalism were gradually loosened and the population did not quickly recover its previous levels, in part because of periodic recurrences of the plague throughout the next 300 years. It was a nation transformed, and the Gough Map, similarly reflecting an uncertain transition into the modern age is an apt mirror of these troubled times.

see more on next page >

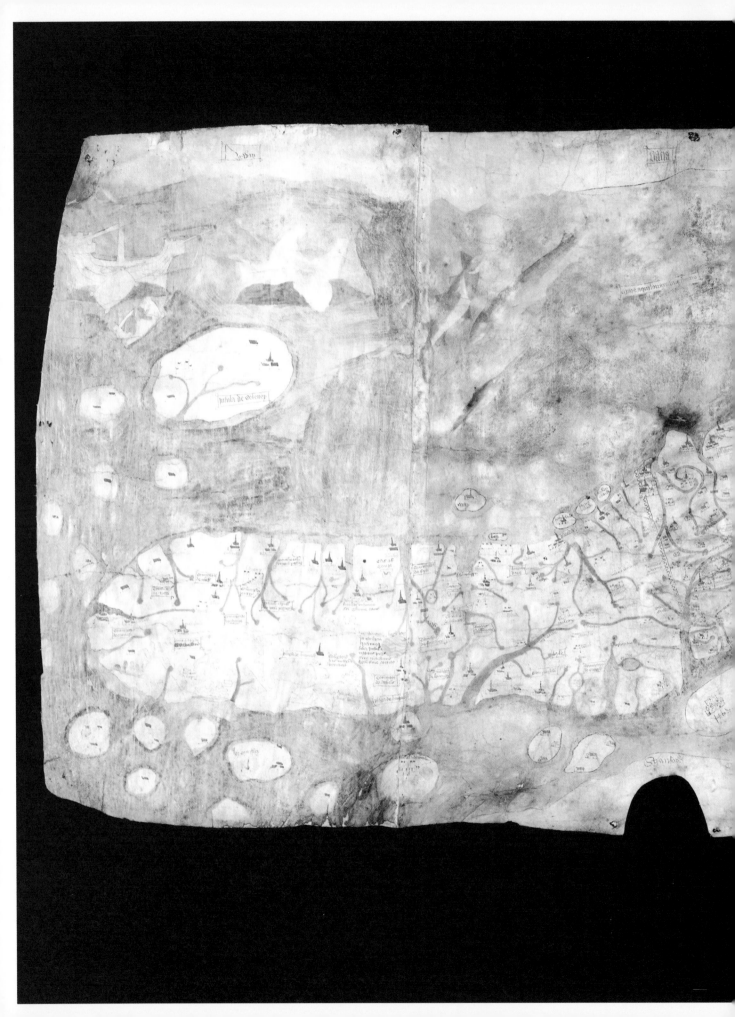

ΩΚΕΑΝΟΣ ΥΠΕΡΒΟΡΕΙΟΣ

ΔΥΤΙΚΟΣ ΩΚΕΑΝΟΣ

ΙΟΥΕΡΝΙΟΣ ΩΚΕΑΝΟΣ

ΙΟΥΕΡΝΙΑ ΝΗΣΟΣ ΒΡΕΤΤΑΝΙΚΗ

ΑΛΒΙΩΝΟΣ ΝΗΣ ΒΡΕΤΤΑΝΙΚΗ

ΔΟΥΝ ΚΟΛΠΟ

ΩΚΕΑΝΟΣ ΟΥΕΡΓΙΟΥΙΟΣ

ΒΡΕΤΤΑΝΙΚ ΩΚΕΑΝΟΣ

PTOLEMAIC MAP OF BRITAIN

Eastern Scotland is twisted at an alarming angle, the Isle of Wight and Orkney magnified unnaturally and the coastline of Britain is curiously jagged in this fifteenth-century map of Britain. Based on the works of the Graeco-Roman geographer Ptolemy of Alexandria, it formed the basis of maps of Britain well into the following century, at a time when some of Britain's national myths were being established.

Claudius Ptolemy lived in Roman-controlled Egypt in the mid-second century AD. A talented astronomer as well as a geographer, he compiled the *Almagest*, in which he espoused a theory of celestial orbs in which the Sun and planets rotated around the Earth, a geocentric theory of the solar system which would remain influential for over a thousand years. In his *Geographia*, a description of the known world, he included co-ordinates for the latitude and longitude of towns and geographical features from which maps could be constructed.

Ptolemy did not compile his own maps – or if he did, they have not survived – but his work was preserved by Arabic scholars, and in the Byzantine Empire, and began to reach Western Europe from the thirteenth century. By the mid-fifteenth century editions including maps – such as this splendid example, emblazoned in Greek – had reached Italy. It was a time when Renaissance scholarship had become fascinated with the revival of classical learning, and as a result Ptolemaic-style maps became the template on which cartographers based their work for much of the next century.

The England which Ptolemy's map depicted with such imprecision was in reality not such an unknown place as far as European dynastic politics was concerned. The country had been embroiled in a war with France since 1337, when Edward III sought to assert his claim to the French throne. In 1413 this claim passed down to a new, young sovereign, as Henry V ascended the throne, aged just twenty-six.

Although it was his ambition to renew the war with France, Henry had several matters to resolve at home. The first was to deal with the Lollards, the followers of John Wycliffe, who made the first translation of the Bible into English and espoused such heretical teachings as believing that the communion wafer was not physically transformed into the body of Christ, and that the Pope should not be the head of the Church. Although Wycliffe had died in 1384, he had a strong following in East Anglia including Sir John Oldcastle, a personal friend of Henry's, who plotted to overthrow the king by arriving at Eltham Palace with an armed party dressed as mummers and kidnapping the royal family. The plot, however, was betrayed by spies and many Lollards were arrested. Henry also saw off another conspiracy by Richard, Earl of Cambridge, to place his cousin, the Earl of March, on the throne, and so, by early 1415 he was in a position to prosecute a war.

It was an auspicious time. The Duke of Burgundy was at odds with the French crown, and Charles VI of France had descended into bouts of madness brought on by the sound of a lance striking a helmet of one of his party while he was out hunting. Much of the time he fancied himself made of glass and could offer no leadership. Even so, it took time for Henry to raise an army of 9–12,000 men, three-quarters of them archers, and to transport them in 1,500 small ships across the Channel. By 12 August he had set sail, beginning the campaign by besieging and capturing Harfleur on 22 September.

From there Henry made what should have been a fatal error, ignoring the warnings of his council and striking directly for Calais. The French army was thus able to intercept him near the village of Agincourt on 25 October 1415, St Crispin's Day. Yet Henry's aggressiveness won the day. He advanced close enough for his longbowmen to rain arrows down on the French, provoking them to charge on a narrow front through marshy ground where they became bogged down, tangled with each other and were cut to pieces by the English men-at-arms. The French lost the flower of their chivalry: 90 counts, 1,500 knights and 4–5,000 men at arms perished and the victory was only marred when Henry ordered the killing of prisoners held in the English camp when he feared the French were about to storm it.

Henry's immediate reward was the conquest of much of Normandy by 1419 and his recognition, by the Treaty of Troyes the following year, as the heir to the French throne. Yet more than this, he achieved immortality as a great martial warrior in the favoured English narrative of history which saw Agincourt as a defining moment of national triumph. That the English had lost almost all their territory in France by 1453, when the Hundred Year's War finally ended, was conveniently forgotten. Just as Ptolemy's influence distorted the outline of Britain, so the need for national heroes distorted the memory of the war against France.

MAP OF INCLESMOOR

This map of the Inclesmoor, near Goole, was drawn around 1407 and shows once more the power of maps to establish ownership in a period of political instability during which, without legal backing, rights could be encroached upon and the interests of the powerful overwhelm those without strong secular or ecclesiastical backing. Its strong colours and almost painting-like quality almost jar against the reason for its drawing; to resolve a hum-drum dispute over the right to cut peat.

The rivers surrounding the moor, the Trent, Don and Ouse are boldly painted, their waters shown swirling where they join the Humber confluence. Even the different roofing of village houses is depicted; some thatched, some tiled. The scene is Inclesmoor, a marshy stretch on the borders of Yorkshire and Lincolnshire. Here wealthy landowners, barons and abbots, had been granted over time the right to dig turves of peat for sale as fuel. In the early fourteenth century the Abbey of St Mary's at York was granted a large section of Inclesmoor for peat-digging. Unfortunately, the boundaries were not well-defined. This led to a dispute with the powerful Duchy of Lancaster, and the drawing up of this map, and a smaller, black-and-white version, as a means of asserting the Abbey's claim to the moor.

Powerful though the abbot was (having his own seat in parliament and being sufficiently wealthy to provide loans to the crown), a similar struggle between York and Lancaster would soon play out at a much higher, dynastic level, and ultimately spell ruin for the Abbey. Henry IV, under whom the charter was drawn up derived his claim to the throne from John of Gaunt, the third son of Edward III. There was, however, an alternative set of claimants in opposition to this Lancastrian branch of the Plantagenets. The Yorkists descended from Edmund of Langley, the fourth son of Edward, and bolstered their claim by an additional connection through Anne Mortimer, the great granddaughter of Lionel, his second son.

The Yorkists, headed by Richard, Duke of York, represented a powerful faction at court, and when Richard returned in 1452 from a posting as Lieutenant of Ireland, he was able to have Edmund Beaufort, the Duke of Somerset removed and, the next year, when Henry VI descended into the first of several bouts of madness, became head of a Regency Council to govern in his stead. Had Henry not recovered, matters might have rested there, but the king regained his mental equilibrium and Richard, by now Lord Protector, was summarily removed.

The result was a crescendo of protest, aggravated by the hardline stance of Henry's queen, Margaret of Anjou, which resulted in a polarization of opinion between Yorkists and Lancastrians and the outbreak of open warfare at the Battle of St Albans in 1455. The Wars of the Roses, named for the white rose emblem of the Yorkists and the red rose of the Lancastrians, would last for almost 30 years. Although Richard was killed at the Battle of Wakefield in December 1460, his eighteen-year-old son Edward survived to become king (as Edward IV) in 1461. Even then the war was far from over, as in 1470 Richard Neville, the powerful Earl of Warwick (known as the 'Kingmaker' for his power to shape events) turned coat and defected to the Lancastrians. Edward was forced briefly to flee to Burgundy before returning the next year and comprehensively defeating the Lancastrians at Tewkesbury on 4 May 1471.

The Yorkists' tenure on the throne was brief. After Edward IV died in 1483, his brother Richard usurped the throne (and most probably had his teenage nephew Edward V murdered in the Tower of London). Disenchantment with his rule and fear of a purge of both supporters of Edward IV and remaining Lancastrian partisans gave the Lancastrian claimant to the throne, Henry Tudor, an opportunity. Setting sail from Harfleur with a party of disaffected noblemen and a squadron of French mercenaries, Henry landed in Wales at Milford Haven on 7 August 1485. Having gathered more retainers, he marched east, and defeated and killed Richard at Bosworth Field in Leicestershire (thanks to a timely defection by the Stanleys whose large force had hovered on the edge of the battlefield waiting to see who looked like winning).

York had lost and it would be Henry Tudor's son, Henry VIII, who ensured that the Abbey of St Mary's, York, lost too. Despite its venerable history, founded in 1055 under Edward the Confessor, and re-established in 1088 by the Conqueror's son, William Rufus, the Abbey had no legal backers, and no powerful protectors when Thomas Cromwell's commissioners came calling. It was one of the grandest abbeys, and so one of the last to fall during the Dissolution of the Monasteries. On 26 November 1539, the Abbot, William Thornton, surrendered its keys and he, and the remaining fifty monks, were pensioned off and the crown acquired the abbey's revenues of £2,085 a year. It all made a dispute over peat turves seem petty.

see more on next page >

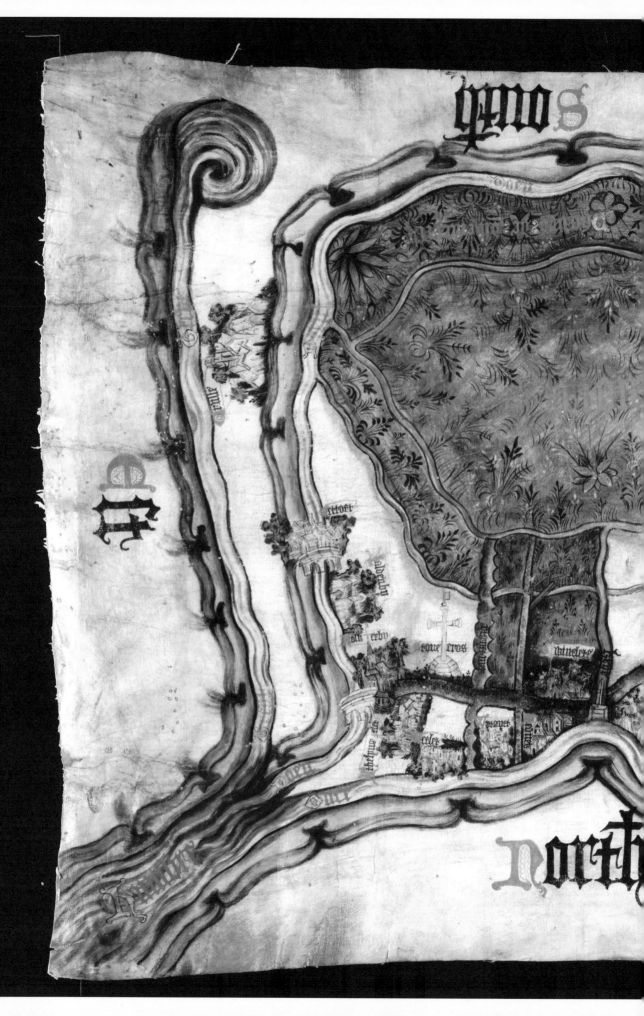

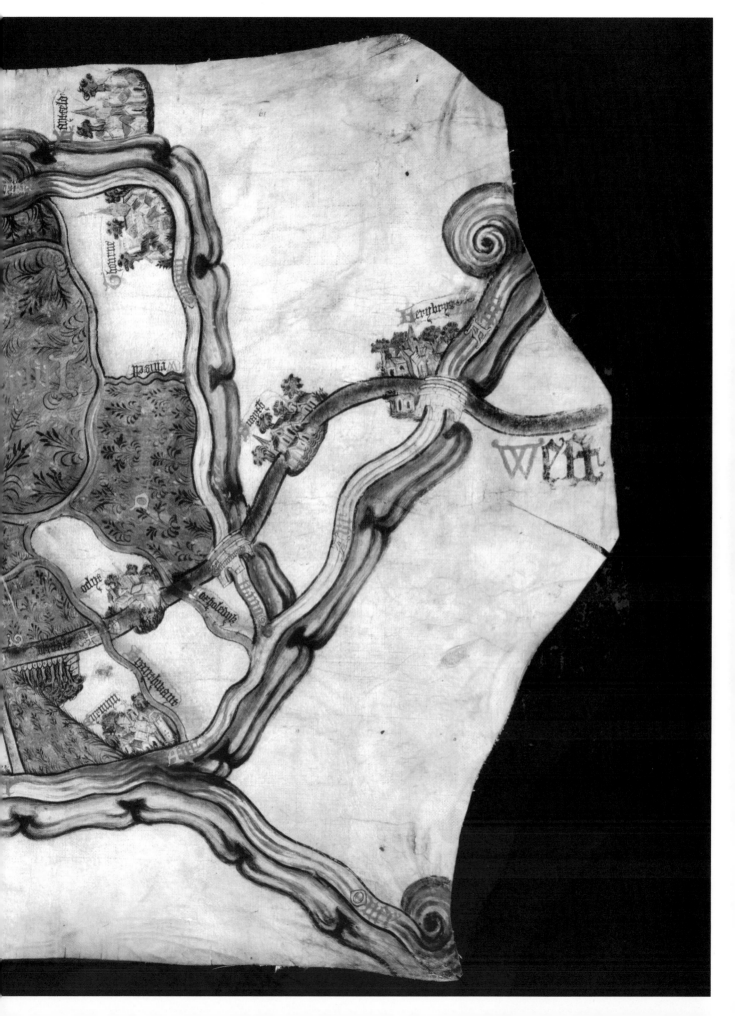

MAP OF THE ISLE OF THANET, Thomas Elmham

The Isle of Thanet is punctuated with churches and other ecclesiastical buildings in this early 15th-century map by the monastic chronicler Thomas Elmham. Yet its pleasing simplicity, uncluttered by other signs save a clump of trees, some crosses, boats and a deer, conceal a more serious purpose – the establishment of the extent of the rights of the church at Minster in a period of profound political instability.

Elmham was a Benedictine monk at Canterbury, who wrote a history of the monastery to which, with studied artifice, this map was appended. The irregular green line which snakes across the centre marks the course of the running deer on which, tradition had it, the borders of Minster's land was based. Thanet was a land steeped in traditions and had played an important role as a gateway to England for invaders and missionaries: it had been the main port of landing for the Roman invasion under Claudius in AD 43 and is said by Bede to have been the spot where St Augustine landed in 597 on his way to bring Christianity to the pagan Anglo-Saxons. It was also the haven the Vikings chose for their first overwintering in 850, marking the start of the intense period of their raids which almost overwhelmed England in the 860s and 870s.

By the late fourteenth century all of this was in the distant past, but ominous political clouds were gathering. Richard II was only 10 at the time of his accession in 1377. The unpopularity of his uncle John of Gaunt, who might have provided strong leadership, together with the faltering progress of the long war with France and the related burden of taxation, meant that discontent seethed and rival factions at court proliferated. In 1381 the imposition of a poll tax led to a shattering outburst of rural discontent as thousands of rebels from Kent and Essex marched on London under Wat Tyler and John Ball, in the Peasants' Revolt. They occupied the Tower of London, where they executed the chancellor, Archbishop Sudbury, before being persuaded by Richard to stand down (Wat Tyler was killed at this meeting, after which his supporters were mercilessly hunted down, tried and put to death).

Richard, now an adult, still struggled to gain control. The Merciless Parliament of 1388 had eight of his leading supporters executed, leaving him under the thumb of the Lords Appellant, his bitter opponents, for the next decade. Peace with France in 1397, by which he married Isabelle, the seven-year old daughter of Charles VI, finally gave Richard the chance for revenge. He had the leading Lord Appellant, Richard, Duke of Gloucester, murdered in Calais, and Henry Bolingbroke, Duke of Lancaster

exiled. In the end the move backfired, as Bolingbroke snuck back to England when Richard was away campaigning in Ireland in 1399, raised enough support from nobles who were fearful that the king's capricious temperament would see them, too, on the scaffold, and seized the throne for himself.

Richard was arrested at Flint Castle and later murdered, but the country was no more stable. Henry IV (as Bolingbroke became) had to face a series of magnate rebellions (in 1400, 1403, 1405 and 1408), an almost empty treasury and a near-permanent state of acrimony with parliament. Worse still, the rebellion of Owain Glyndwr, which swept across Wales from 1400, almost succeeded in driving the English out. The self-declared Prince of Wales held parliaments at Harlech and Machynlleth and seemed for a while to have united Welsh sentiment behind him. But once Henry was able to concentrate on Wales, the rebellion was throttled – Harlech and Aberystwyth were retaken in 1408–9 and Glyndwr ended as a hunted fugitive in the Welsh mountains, whom the English never quite tracked down.

Henry V, who succeeded his father in 1413 was more fortunate. England was quiescent once more, and the young king was able to take advantage of a state of near civil war in France, ignited when Charles VI went mad. Exploiting a strategic alliance with Burgundy, Henry launched a major invasion of France, an expedition on which Thomas Elmham accompanied the army as royal chaplain. The monk was present on the field of Agincourt on 25 October 1415 when Henry's longbowmen scythed down ranks of French knights who were forced to charge across a narrow frontage through boggy ground. France was prostrate and, after Henry had seized Normandy in 1419, was forced to sue for a peace at Troyes in 1420 which gifted the English king most of the north of the country.

Elmham may have though he had lived to see stability return, but it did not last. Within forty years the English had been driven out of all of France save Calais, undone by the inspirational example of Joan of Arc who exhorted the French to lift the English siege of Orléans in 1429. And England itself descended into civil war in the 1450s, when partisans of Richard, Duke of York, a descendent of Edward III fought with supporters of the Lancastrian claimant Henry VI in the Wars of the Roses. Amid such turmoil, land claims could be overturned and old-established rights ignored. It was as well, in such times, to have the support of an ancient charter or map, even if it was based on the wanderings of a deer.

Mappa Thaneti Infule

Oriens

Sci Petri

Mergate

Sci Iohannis   Salmeſtone

Sci Lauretii   Stanore

Aquilo   heling dene Wallis

Aldelond   Auſter

Iſtud moſteriū fundavit Edburga Abbatiſsa poſt mortem S.dæ Mildredæ

Wode cherche   Aldelonde

Curſus cerve

LIBERT. BARONIA   Monſtru Ator Pet et Pauli

Tene   Kingeſto

Puteus thunor   Iſtud Moneſteriū fundavit Domneva

Berchington   Ecclo. S.t Maria

Clyve   Menſtre

Parker   Plumſted hanc

Eldricke radulphi   hoo Reye tey   Boxile

Omū Scōrum   Screves Kope

Sci Nicholai   Docnerd hope

Monketon.

de domio Baroniæ

Sci Egidii.

Occidês

Saerre

Reculuere

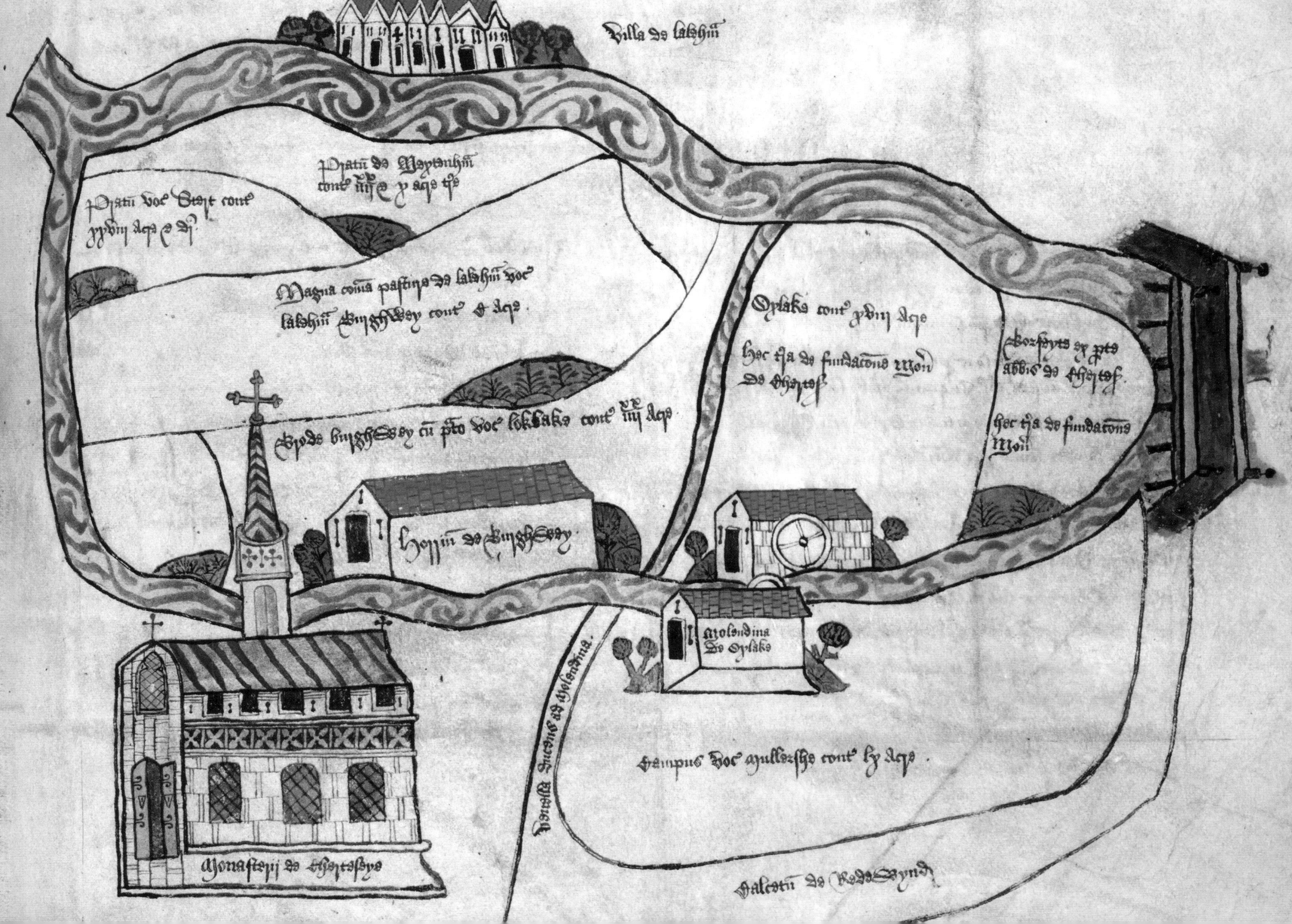
Villa de Lalehm
Oxlake cont' xviij acr'
Hec t'ra de fundacone Mon' de Chertesey
Molendina de Oxlake
Campus voc' Mullersshe cont' xl acr'
Monasterij de Chertesey

MAP FROM CHERTSEY CARTULARY

The colours are startlingly vibrant in this fifteenth-century plan from the cartulary (book of charters) of the Benedictine abbey of Chertsey in Surrey. Originally drawn up to illustrate the monastic case in a dispute with the nearby village of Laleham over grazing rights in the abbey's pastures, it was preserved as a way of proving Chertsey's rights in any future legal cases; an ability the monastery, as an island of learning in an age when literacy was comparatively limited, was in a privileged position to exercise.

Chertsey had a venerable history. It was, tradition related, founded in AD 666 by St Erkenwald and acquired considerable lands to the south of London, initially under the patronage of Mercian kings such as Offa (757–96), before ending up in the territory of the kingdom of Wessex. The abbey's history nearly came to a premature end when it was sacked by Danish Vikings in the late ninth century and Abbot Beocca was martyred (and later canonized), but after a century or so in abeyance, the house was refounded in 964 by monks sent from the abbey of Abingdon.

By the time of the Domesday Book, William the Conqueror's great summary of all the landholdings in England compiled in 1086, Chertsey's landholdings had become considerable again, though over time some was sold and the monks' husbandry of what remained was not assiduous, so that it took a considerable effort under Abbot John de Rutherwyk (1307–46) to rationalize the estates and to exploit them more efficiently.

All this required accurate record-keeping. It was the monks' secret weapon in their struggle with neighbours and secular lords; their charters had an air of sacred authority and, with their libraries and scriptoria, where skilled monks copied manuscripts by hand (the only way of disseminating or duplicating them in the days before printing), they almost always had the upper hand in being able to produce accurate records (and the odd well-targeted forgery) from days gone by.

So it was with the dispute which this map illustrates. The abbey is boldly drawn at the bottom of the map, with its towering spire, and nearby, the abbey's mill, with its water wheel, while the field boundaries, lanes and Laleham village (at the top) are all sketched out in a bright, clear palette demonstrating Chertsey's right to mastery over the 'Island of Burway', lying between two branches of the Thames.

Through the vicissitudes of the Wars of the Roses and the accession of the Tudors, the abbey must have thought its patrimony well-cared for and safe for future generations of monks. For a time, it was even the site of a royal burial, as Henry VI's body rested there from his death in 1471 until Richard III ordered its removal in 1484 to St George's Chapel in Windsor, as a means of vicariously associating himself (a usurper) with a rightful king (albeit a Lancastrian, and one he and his Yorkist family had fought a long war against).

Then, in 1537 disaster struck. Short of money and finding himself at the head of a Church of England now sundered from its loyalty to the Pope in Rome, Henry VIII ordered commissioners to inspect the monasteries of England, and then to dissolve them. The process was carried out with ruthless efficiency by agents of the chancellor, Thomas Cromwell, who had persuaded his master that they were centres of impiety, where monks roistered and caroused rather than prayed, and from where a great store of silver made its way to Rome.

In July 1537, the commission reached Chertsey. There was little the last abbot, John Cordrey, could do, save acquiesce. On 6 July 1537, together with the prior and thirteen remaining brethren, he surrendered the abbey to the crown. The face-saving gesture of being granted an alternative home at the abbey of Bisham did not last long, for that, too, was dissolved less than a year later. For all the charters lovingly drawn up, and for all the maps drafted as a permanent record of the abbey's right to its lands, its nine centuries of existence came to an abrupt end.

MAP OF SCOTLAND,
John Harding

With its depiction of a country bursting with fine castles, walled towns and impressive churches, John Harding's 1457 map of the Scotland (the first to show it independently, and not as part of a map of Britain or the world) gives an impression of a rich and prosperous realm. Yet its true purpose was not to praise Scotland, but to induce the English king to invade and conquer this enticing prize.

Harding's map accompanies his rhyming Chronicle of English history which he began around 1436 after his retirement from royal service which had seen him fight at Henry V's resounding victories against the French at Agincourt and Harfleur. Harding's retirement, though, was not absolute and in 1440 he was despatched to Scotland to hunt out documents which might help establish the English king's right to the Scottish crown. For his services, he received an annuity of £10 per year. Seventeen years later, Harding went north again and in June 1457 he handed over six documents supporting English sovereignty over Scotland to Henry VI's treasurer, John Talbot, Earl of Shrewsbury, so securing himself a further annual pension of £20. It was at about this time that Harding drew up his map, accompanied by a suggested itinerary and invasion plan for the expedition he urged Henry to launch.

That invasion never came (or at least not for nearly a century, until Henry VIII crossed the border in 1544), but it was not an outlandish prospect. Ever since Edward I had secured the fealty of 2,000 Scottish noblemen recognizing him as King of Scotland in 1296, successive English monarchs had turned their eyes northwards in the hope of absorbing Scotland once and for all. The crowning of Robert Bruce as king at Scone in March 1306, and his crushing defeat of the English at Bannockburn in June 1314, when the Edward II's knights got bogged in marshy ground, and his archers were too constricted to fire without striking their own men, seemed to put an end to English hopes.

Further attempts at meddling by trying to have Robert Bruce's excommunication reconfirmed by Pope John XXII backfired, and led to the Declaration of Arbroath in 1320 at which Scottish notables declared: 'It is in truth not for glory, nor riches, nor honours that we are fighting, but for freedom alone, which no honest man gives up but with life itself.' When the Pope demurred and refused the excommunication, the English backed away, and the Peace of Northampton in 1328, by which

the English recognized Scotland as an independent nation, seemed to recognize the status quo.

Yet Scotland's monarchy was plagued with bad luck. When Bruce died in 1329, his heir, David II was just five years old and, in an attempt to preserve him from an English invasion in 1333, he was sent to France where he spent eight years, undermining Scotland's stability as rivals jostled to dominate the regency. When he returned he answered the call of his French allies to provide them with aid after their defeat by the English at Crécy in 1346, but his invasion that year of northern England was a disaster: he was defeated at Neville's Cross and spent eleven years in exile as a captive of Edward III.

The ransom paid for his release crippled Scotland and contributed to a state of endemic disorder during the twenty-year reign of Robert II, who had succeeded his father in 1371. His son, Robert III had been crippled after being kicked by a horse and for 25 years from his accession in 1390 Scotland was ruled by regents. In 1406, when rumours arose that one of these was about to hand over Robert's son James to the English, the young prince was sent to France, but his vessel was intercepted by pirates and he was passed on to Henry IV and held a hostage for the first 18 years of his reign (his father having died of shock on hearing of his capture).

Although Henry V allowed James I, by now 29, to return home in 1424, secure in the knowledge that his acquisition of half of France by the Treaty of Troyes four years previously had neutered the power of the French. Yet Henry was wrong; the spark of French resistance was fanned by Joan of Arc, and, with renewed vigour, the French king pushed back against the hated English occupation.

It was perhaps in this context that the Harding, himself a veteran of the war against France, thought that Henry might be tempted into solving the Scottish question once and for all in England's favour. And so he drafted his map, a cloth of prosperous-looking towns, sturdy fortresses and plump churches ripe for the picking. Yet by 1453 the English were on the back-foot. Driven out of almost all of France after the Battle of Castillon, they had lost the Hundred Years' War and England was about to descend into a debilitating war between Yorkist and Lancastrian claimants to the throne. Harding's map, and his schemes for invasion, were quietly forgotten.

see more on next page >

Oute Iles

Oute Iles

Catteneffe

Athell

North Se

Murref

Catteneffe

Erne

Bolhaban

Ros

Re de castel

Garyogh

Erbroth

Arbir deue

Mane

and called it Bellyngesgate after his owne name
Reygnud nobly all his lyf and lieth at next Troye

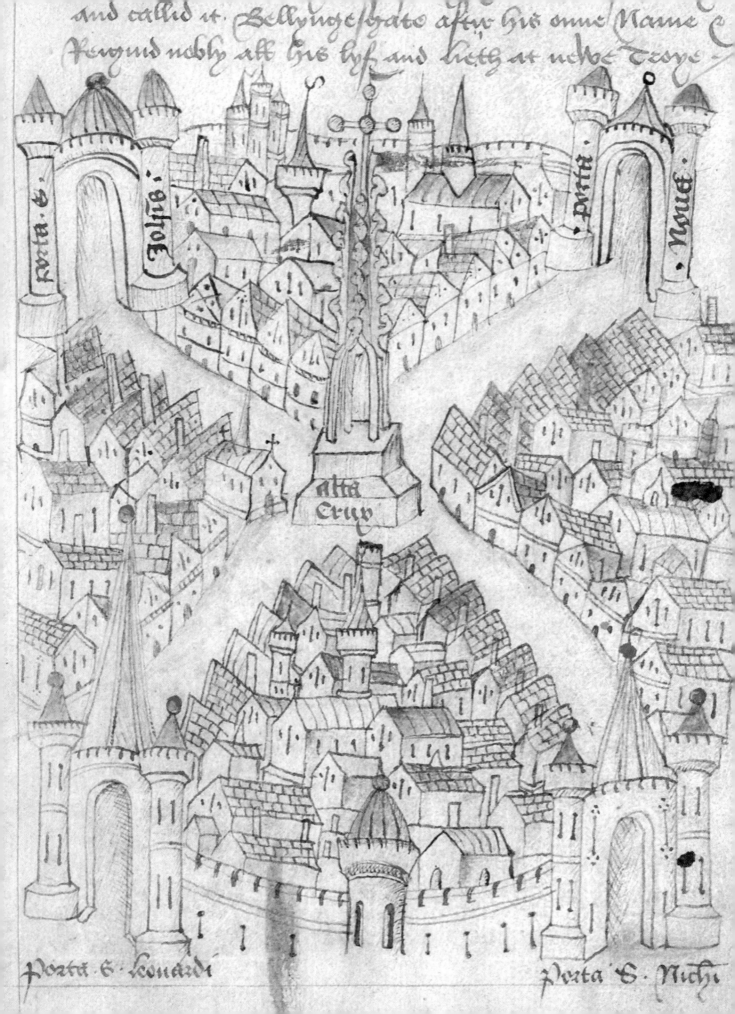

Porta S. Leonardi   Porta S. Nicho.

PLAN OF BRISTOL, Robert Ricard

Robert Ricard's striking bird's eye view of Bristol in around 1479 shows a colourful view of the city radiating out from the central High Cross in the direction of its four main gates. The image of a crowded, prosperous settlement, however, is misleading. Bristol was in the midst of an economic crisis and one which would soon see the launching of a venture to solve it: Britain's first voyage to the New World.

Bristol had been an important port since Anglo-Saxon times and by the thirteenth century was a busy entrepôt, whose merchants exported woollen cloth and tin, bringing back home in return cargoes of wine from Bordeaux. Its wealth led to the granting of a charter by Edward III in 1373, in celebration of which the burghers erected the High Cross in the centre of the city (whose alcoves were ultimately adorned with statues of eight English monarchs, the last being James I). As well as the site of markets and fairs, it also became the favoured venue for public executions, most notably in 1400 that of Thomas le Despenser, Earl of Gloucester, who was beheaded for his complicity in a plot to overthrow Henry IV and restore the deposed Richard II.

Ricard was the clerk to the city council and would have understood the baleful effect of the end of the Hundred Years' War on Bristol. Bordeaux had been lost to the French in 1453 and trade suffered an immediate setback; between 1455 and 1460 the levels of wine imports fell by almost half, while those of cloth tumbled by 40 per cent. Although there had been some recovery by the 1490s, Bristol merchants were soon eagerly seeking new markets to make up the shortfall.

An increase in voyages to Spain, and in particular Portugal, may have led them to hear whispers of lands said to lie westwards across the Atlantic, and in the 1480s several expeditions were launched to find the mythical island of Hy-Brasil, whose fabulous wealth grew with each telling of the tale. Into this swirl of desperate hope and dashed expectations a Venetian merchant arrived in Bristol in 1495.

John Cabot (or Giovanni Caboto) knew how to twist the urgent need for new trade outlets and the fervent belief in Hy-Brasil to his advantage. Crucially, Henry VII happened to be visiting Bristol that winter, and, having heard of the marvellous discoveries made by Cabot's fellow Italian Christopher Columbus, the king was in a receptive mood to any project that enabled England to seize a share in the new lands across the ocean.

Cabot was awarded a royal patent in March 1496 which enjoined him to 'to find, discover and investigate whatsoever islands, countries, regions or provinces of heathens and infidels, in whatsoever part of the world placed, which before this time were unknown to all Christians'. These would, naturally fall to the English crown, in exchange for a healthy royalty payable to Cabot.

The results from the first expedition were paltry. After fifty-two days at sea, nervous lest they were heading into an endless watery void, the crew of Cabot's single ship, the *Matthew* sighted land, probably off the coast of Newfoundland. They found no cities dripping with gold, not even any natives, although they came across the remains of a fire which indicated there were, somewhere, inhabitants. The sole promising encounter was with the enormous shoals of cod they found a little offshore.

Despite a certain sense of disappointment, Cabot talked up the results of the expedition to Henry VII and in early 1498 was given permission to undertake a further voyage. The king was in the midst of fighting off a dangerous rebellion by Perkin Warbeck, a young Fleming who claimed to be Richard, Duke of York (one of the princes murdered by Richard III in the Tower of London in 1485), and could not give Cabot his full attention. Still, he awarded the insistent Italian a grant of £10 and the title of 'Grand Admiral' (which matched the accolade Columbus had received from the Spanish monarchs Ferdinand and Isabella for his discoveries).

In May 1498, Cabot set sail from Bristol with five ships, laden with a cargo of lace to trade with the merchants of the Americas whom he still had hope of finding. Nothing more is known of the voyage, whether Cabot perished or whether the expedition was such a complete failure that it was quietly forgotten. Yet for Bristol it would ultimately open up new horizons; those shoals of cod would provide an ample income for West Country fishermen, and ultimately the trade in tobacco and cotton, and the shipping of slaves from Africa to the plantations of the New World, would eclipse the wealth that had been derived from the Bordeaux wine trade.

PLAN OF FRENCH RAID ON BRIGHTON

Plumes of smoke rise from the village of Brighton (or Brighthelmstone as it was known in Tudor times) in this depiction of a French raid on England's south coast in 1514. The near-destruction of the settlement was the price for the young Henry VIII's romantic thirst for chivalric glory, which had gravely angered the French.

The map, drawn about three decades after the event by the court draughtsman Anthony, is more a cautionary tale than a precise depiction of that early summer's day when the French raiders landed. The ships which buzz around the harbour like hornets are more appropriate for a time around 1542, when another war scare swept the English court and it was feared the French would return.

Henry had been itching for a war with France almost since his accession in 1509, but the formation of the Holy League between Venice, the Papacy, the Holy Roman Empire and Spain offered him an opportunity, aimed as it was at curbing French ambitions in Italy. Having raised the requisite funds from parliaments in 1510 and 1512 – never easy moments for the irascible Henry as the members of parliament almost always sought concessions in return – the king sent an expeditionary force to Gascony to join forces with Ferdinand of Aragon (Henry's father-in-law). The campaign was a disaster as the 12,000 men under the Marquess of Dorset were supposed to attack Bayonne, but Ferdinand instead attacked Navarre, far to the east and left the English force stranded.

The next year Henry took to the field himself, with a large, lumbering force that rolled slowly towards its target of Thérouanne. The bloated baggage train and pedestrian pace of the column gave the French ample time to prepare, and it took from June to August to cover the ground to begin the siege. The English, however, were surprisingly successful and, after a field skirmish, following which Henry knighted dozens of men so that it came to be called the Battle of the Spurs, Thérouanne surrendered on 24 August.

Gratified, Henry returned, but the new English outpost and a series of raids on the French north coast stung the French into action. A fleet was assembled under Admiral Pregent de Bideaux and ordered to teach the English a lesson. It was this force that descended on Brighton, catching the village completely by surprise. Before reinforcements could be summoned from Lewes, almost all its buildings save the church of St Nicholas had been set alight. As the English counter-attacked a rain of arrows descended on the French, one of them striking Pregent (referred to on the map and the English accounts of the battle as 'Prior John') in the eye. Now under pressure the French re-embarked and returned across the Channel. So relieved at his survival was Pregent that he later donated a wax model of his face, complete with arrow protruding from it, to a church in Boulogne.

Henry was determined to fight another glorious campaign but for the next decade his Machiavellian chancellor, Cardinal Wolsey, managed to steer him through the treacherous waters of European diplomacy without definitively opting for either a French or an Imperial alliance. In June 1520 Henry met with Francis I of France at a rendezvous near Calais so lavishly decked out that it became known as the Field of the Cloth of Gold. It took 6,000 workmen to set up the tents and pavilions necessary to accommodate the host of English and French nobility that attended. The cooks were kept busy, too, with menus which called for the importing of 700 conger eels and 26 dozen herons to satisfy the appetites of the royal courts.

Henry and Francis jousted almost every day for over two weeks, on one occasion the French king landing his Tudor counterpart flat on his back. It was all high adventure for Henry's chivalric sensibility, but diplomatically it meant nothing. Late the following year, Henry signed a treaty with the Emperor Charles V and in 1522 the Earl of Surrey began a new series of raids against northern France. Only the king's 'Great Matter' – his determined campaign from 1527 to secure a divorce from Catherine of Aragon in order to marry his new paramour Anne Boleyn – kept him from further military adventures on the continent for over a decade. Not until 1542 did war seem to be imminent again, and when it did, the memory of the raid on Brighton in 1514 served as a reminder that the English south coast remained ever vulnerable.

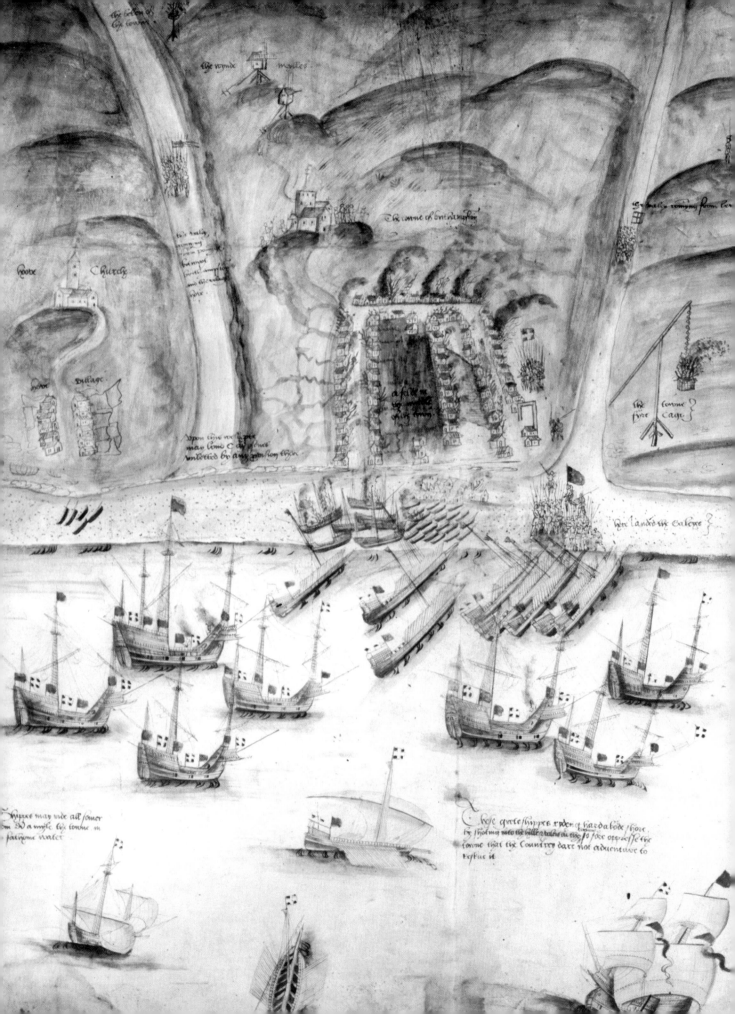

A Plott for the making the Hauen of Douer.

PLAN OF DOVER HARBOUR,
Vincenzo Volpe

Lying almost at England's closest point to France across the English Channel, Dover was long a key target for any invading force. In the 1530s concerns began to grow about its defence and this 1532 watercolour represents one of a series of schemes to improve the town's harbours and fortifications at a time when England's Tudor monarchs were beginning to harness the power of cartography to improve their knowledge and administration of their realm.

Maps, which had long been the province of scholars or local administrators, came increasingly to the attention of the court from the 1520s, as a means of understanding the topography of the provinces (and to defend, and tax, them more efficiently). With this in mind, in 1530 the government commissioned the Neapolitan painter Vincenzo Volpe to create maps of the strategically important ports of Rye and Hastings. He clearly made an impression, and in 1532 the Corporation of Dover asked him to draw up a proposal for a new harbour and fortifications for the town, which was becoming an urgent necessity, both because of the threat from France, and the more pressing damage caused to shipping by storms and to navigation by the silting up of the existing inadequate anchorage.

The resulting plan is clearly more the work of an artist than a surveyor, evoking the town and the surrounding landscape more powerfully than it lays out the precise details of the new defences. The vision of Henry VIII was similarly unclear, and in a sense romantic. After he ascended the throne in 1509 he became entangled in an expensive diplomatic dance with Francis I of France and the Holy Roman Emperor Maximilian I, hoping to play off one against the other, but ending up, by 1513 having almost emptied the Exchequer with the expenditure of a million pounds. A minor victory that year against the French at the Battle of the Spurs was exaggerated by Henry, who gave the encounter its name by knighting dozens of men for their heroic deeds, and his then ally Maximilian, who published a joint account of their glorious doings on the battlefield.

Chastened by the negative reaction of parliament to his attempts to raise money for further expeditions and by the near revolt that broke out in Norfolk and Kent in 1522 in response to a compulsory loan he had ordered (euphemistically named the 'Amicable Grant'), Henry pulled back from foreign military ventures. In any case, he was too preoccupied for the next six years with his attempt to rid himself of his wife, Catherine of Aragon, who had failed to provide him a male heir. That she was also the aunt of the new Holy Roman Emperor, Charles V, meant that Henry had a powerful set of opponents, including the Pope, arrayed against his demands to have his marriage annulled.

Henry finally engineered the divorce, by disposing in 1529 of Cardinal Wolsey, his chief minister, whose religious scruples made him too loyal to Rome, and replacing him with the much more ambitious Thomas Cromwell, who pushed the matter through. In 1534 he had himself declared the head of the Church in England by the Act of Supremacy, beginning the process of a definitive break with the Catholic Church (which would, in 1536 handily solve his revenue crisis when the dissolution of the monasteries released an enormous rush of funds into the Exchequer).

While it gave him increased power over religious affairs in England and allowed him to marry his mistress Anne Boleyn (and ultimately to get the male heir he craved, when the future Edward VI was born in 1537 to his third wife, Jane Seymour), the King's 'Great Matter' also made him enemies. A Papal Bull of 1535 called upon all Christians to attack the English king, and neither Francis I or Charles V, as good Catholic monarchs, could readily treat with the excommunicated Henry.

Although England had been at peace with France since 1525, the need to attend to the defence of the south coast ports was now all the more urgent in case the French saw fit to carry out the Pope's commands. As a result, a series of further proposals to improve the fortifications of Dover was put forward. None of them came to fruition, however until the 1580s, when a scheme proposed by Thomas Digge was carried out. That he was a mathematician, and his father a surveyor, perhaps proved that the artistic scheme of Vincenzo Volpe was as doomed as the king's idea that he could swiftly and painlessly dispense with his wife.

see more on next page >

19.6

A Plott

the Hauen of Dover.

COASTAL DEFENCE MAP,
prepared for Thomas Cromwell

In 1539 England was swept by an invasion scare, as fears spread that a rapprochement between Francis I of France and the Holy Roman Emperor Charles V might lead to a Catholic alliance to avenge Henry VIII's break with Rome. This map, a scroll 3.5 m (11½ ft) long, was the result of a frantic surveying effort to identify possible weaknesses in the defences of England's south coast and to order new fortifications to plug gaps through which the French and Imperial forces might surge.

In February 1539, the king's chief minister Thomas Cromwell ordered surveyors 'of every shire in Ingland being nere the see . . . to viewe all the palces alongest the cost where any daungers of invasions ys . . . and to certifie the sayd daungers and also best advise for the fortificacion thereof'. The task was vital, for war seemed imminent. Francis I and Charles V had just reached an agreement at the Treaty of Toledo to end the conflict between them, and so they were now free to turn on the heretic (and excommunicated) Henry. The information Cromwell's men sent back to Greenwich was compiled into a great map of the south coast from Land's End to the River Exe. It emphasizes the beaches, as though highlighting the areas where an enemy might land, and documents in great detail fortified points, adding in proposals for new towers and forts, which are labelled either as 'not made' or 'half made'.

The preoccupation led to the strengthening of defences elsewhere in England; the fortress-building programme extended as far north as Berwick and men such as the Italian engineer Archangelo Arcano designed Italian-style fortifications (or *trace italienne*) with angular towers and strongpoints which enhanced the power of the defenders to inflict raking shots on anyone trying to storm their ramparts. The final cost of the cartography alone was prodigious – amounting to £376,000 by 1547, which was more than Henry spent on his palaces.

Henry also had good reason for concerns domestically. The religious changes he imposed in the 1530s had been resented by Catholic traditionalists, and the stronger royal bureaucracy championed by Thomas Cromwell rankled with conservative nobility in the provinces. In 1536 Yorkshire erupted in the Pilgrimage of Grace, a rebellion which mixed religion, feudal sentiment and economic grievances in equal measure. It was brutally put down by 1537, and hundreds of its ringleaders and many comparatively innocent participants were hanged or burnt. The rebellion provided the pretext for the dissolution of the larger monastic houses and abbeys (previously only those with an income of less than £200 had been closed down), which in turn provided an influx of funds for Henry, allowing him both to carry out his programme of fortifications and to renew his foreign military adventures.

An initial attempt to attract German allies, and to sidestep the Emperor, came to grief when a marriage alliance brokered with the Duchy of Cleves in 1540 proved not to Henry's taste. The king unkindly referred to Anne of Cleves as a 'Flanders mare' and claimed he had been misled by the flattering portrait of her sent over by the court painter Hans Holbein. Nonetheless the ailing Henry, now crippled by infected leg ulcers, wanted to prove his continued martial prowess (his sense of his own youthful vigour already confirmed by his marriage to his fifth wife, the teenage Catherine Howard, later in 1540). He launched an invasion of Scotland in 1542, and two years later turned to the old adversary, France. Henry was so ill that he had to be carried into Calais on a litter in July 1544. Even so, the royal army managed to provide one last victory and in October captured Boulogne. The king had boastfully proclaimed that he would march on Paris, but the expense of the war was becoming as ruinous as his own ill-health. With a bill for £750,000, which meant that most of the bounty of the dissolution of the monasteries went straight into the hands of military suppliers, he turned home to eke out the last two years of his life in growing physical pain.

By the time Henry died in January 1547, the threat of war had receded. England instead turned inward, as it experienced first a wave of radical Protestantism under the boy-king Edward VI and then a Catholic reaction under his elder daughter Mary. Cromwell's map, though, was a warning to all that the danger from the south had not entirely disappeared.

see more on next page >

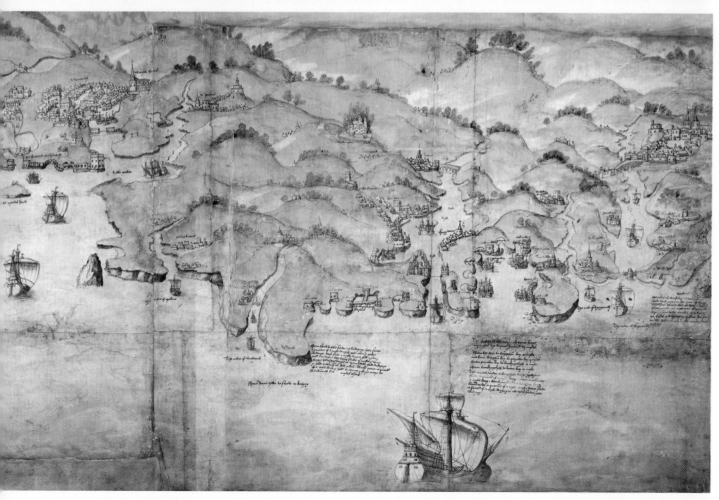

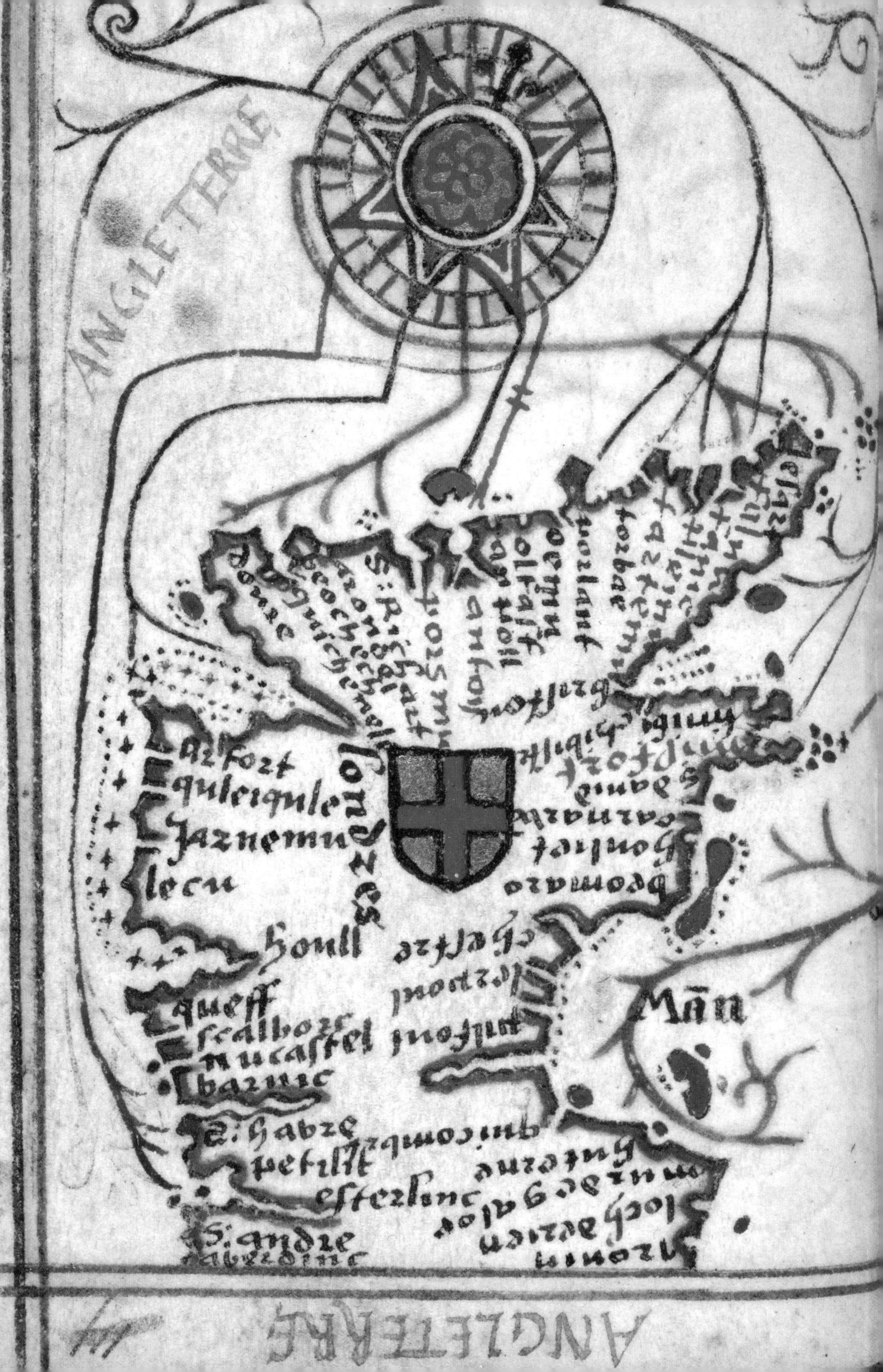

al fort
guleigule
Jaznemus
lecu

Londres

houll   Chesse
Perton

queff
scalbore
nucastel motpul
barme

a: habre
petill   guicomeric
esterlinc   auraing
Gutorne

S. andie   miae galor
lochaaien
Mān

S: R: ohart
georchichenol
donge

norland
oliaiff
anborp

fazberme
tilstre
anfort

finib bibith
frotfort
laus
caruaue
fontet
bromato

 S: R: ohart

TIDAL CHART OF SOUTHERN BRITAIN, Guillaume Brouscon

The chart of England's south coast produced in 1540 by the Breton cartographer Guillaume Brouscon shows the increasing interest in naval mapping of Britain's shorelines, of particular importance to European merchants wishing to trade there. Yet it also highlights the country's vulnerability; in order to take advantage of the security its island status in theory offered, the English crown was forced to begin to build up a strong naval force.

Brouscon's chart comes from his *Almanachs*, a compilation of naval maps produced at a time when interest in rutters, or seamen's guides, was growing. He confines himself to coastal detail – there is almost nothing of note included inland – and important ports (such as Portsmouth) are picked out in red lettering. A particular novelty was his inclusion of cadrans, a type of compass diagram from which lines snake out, indicating the time of high water at a port on the day of the new moon. By adapting this calculation, mariners were able to work out the window within which they could easily enter a port.

The publication of Brouscon's tidal chart came at a time of particular tension between England and France (which would flare up into outright warfare, including naval raiding, in 1542). Naval defence was not a new concern and as long ago as 896 King Alfred had constructed a small fleet to defend against the depredations of seaborne Danish Vikings, but the commitment of subsequent English kings to sovereignty of the sea had been distinctly intermittent. During the Middle Ages, large fleets were assembled, such as the flotilla of 700 vessels gathered by Edward III in 1347 for the war against France. But these were ad hoc affairs, and soon dispersed once the need was over. Even when large ships were specifically built for royal use, such as Henry V's *Grace Dieu*, at 1,400 tons a medieval naval leviathan, they were left to moulder in port once the war subsided.

Henry VII began a restoration of the navy, ordering 'Great Ships' such as the 450-ton *Sovereign* and 600-ton *Regent* and overseeing the construction of England's first dry-dock at Portsmouth in 1495, but the real naval renaissance happened under his son, Henry VIII. With a flourish of youthful energy and not a little hubris, Henry ordered the building of newer and larger ships at the start of his reign, including the *Henry Grace à Dieu*, a vanity project of 1,500 tons which saw little action. A more mature phase started in the 1530s, when under the guidance of Thomas Cromwell, the king began to shape a more efficient navy, which ultimately included fifty-three ships. The fleet even saw genuine action against the French during the Third Anglo-French War of 1542–6 when several naval battles in the Channel showed that the investment in the Royal Navy had been a wise precaution.

Henry also established dockyards on the Thames at Woolwich and Deptford and in 1546 set up a Navy Board, consisting of a Treasurer, Controller, Surveyor of Ships, and Master of the Ordnance. Yet after his death in 1547, the impetus faded and by 1558 the fleet had declined to just twenty-four ships in number (and its tonnage had dropped from 11,000 to 7,000). It would take a long process of reconstruction before England's navy was strong enough to face its next major challenge, the Spanish Armada of 1588.

CHART OF THE SOUTH COAST, Jean Rotz

This chart of the England's south coast (and the north coast of France), probably by the Dieppe master Jean Rotz and dating around 1542–4, shows the mapmaker's concern for accurate portrayal of the coastline, with the topography, especially the beaches and cliffs, almost lovingly detailed. Allied to this is the mariner's preoccupation with the winds, shown as anthropomorphized human heads blowing from the relevant direction. Ironically, at a time of almost poisonous relations between England and France, all this was drafted by a Frenchman in the service of Henry VIII of England.

Amid the delicate penmanship is a sign of the times. The death's head that does service for the winds blowing from the south (and thus from France) is perhaps a comment on the looming war between the two countries. Rotz was a recent arrival (although his family was originally Scottish) and his position in England was precarious, a situation he perhaps tried to make good by donating the king his recent works on hydrography and a treatise on the compass.

Henry, though, was in the mood for war, and that foreign mapmakers might be of use in conducting this was no cause for royal concern. His marriage to Anne of Cleves in 1540, conceived as a means of gaining German allies to circumvent the wearisome diplomatic dance with Emperor Charles V and Henri II of France, had failed disastrously. When Henry finally saw the bride-to-be – the match had been made at a distance, encouraged by a flattering portrait of her by Hans Holbein – he confessed 'I am ashamed that men have so praised her as they have done – and I like it not' and almost immediately set off divorce proceedings. Anne's replacement, the teenage Catherine Howard, turned out to be a libidinous liability and once her many affairs came to light, the king was forced to have her executed in February 1542.

Henry distracted himself from his marital woes with an expedition against Scotland, and once he had thought it won, in 1543 he declared war on France. A desultory sally against Flanders was followed by a full-on invasion launched in July 1544 from the English enclave of Calais. Henry, excited to be recapturing the martial glory of his youth, insisted on accompanying the army but, in reality ageing and infirm, he had to be carried around on a litter. The English generals, the Dukes of Norfolk and Suffolk, were equally lacking in youthful vigour and, over-cautious, they failed to thrust towards Paris, contenting themselves with the capture of Boulogne.

Still, it was a prize worth having and Charles V, Henry's notional ally, fell out with the English by insisting on a share. The French, thwarted on land, turned to the sea to win back their pride. In May 1545 they mustered a large fleet on the Seine, intending to launch an invasion. Their 80 ships faced a flotilla of 128 commanded by Viscount Lisle which, on sighting the French on 16 July instantly took to the shelter of Portsmouth harbour. The pace of the battle was leisurely at first, with a light cannonade by the French against the English lines being the only action at note. On the night of 18 July, Henry VIII even dined aboard Lisle's flagship, the *Henry Grace à Dieu*, so slight was the danger considered to be. Yet the next day, the wind dropped and the English fleet, immobile, found itself a sitting duck for the French cannons. The *Mary Rose*, a 700-ton carrack and the vice-admiral's flagship, intervened, but, whether because it turned too fast into the wind, or fired all its cannon at once and overbalanced is not clear, the ship turned turtle and sank. Hundreds died, and only thirty sailors swam clear of the wreck.

It was the biggest naval catastrophe of Henry's reign, but the French were unable to take advantage. A landing on the Isle of Wight on 21 July was beaten off by local militia and, running short of supplies and with his own ship leaking, the French admiral Claude d'Annebault paused for one last plundering raid on Seaford in Sussex before heading for home. The expedition was larger than the Spanish Armada of 1588 and had been an utter failure.

Not all was lost, however. Henry VIII was tired of the war, which was proving hugely expensive and a treaty was concluded at Ardres in 1546 by which the English were to retain Boulogne, though the French would be permitted to buy it back in 1554. In the event, the war sparked back into life in 1548 and two years later, with its defence proving too costly, Boulogne was simply handed over to France. The war had cost Henry VIII over £2,000,000 and for little gain; financing it forced the alienation of £800,000 of land, costing the crown most of the gain it had made by the dissolution of the monasteries.

Jean Rotz, meanwhile, sensing the antipathy to emigrés that was growing after the death of Henry VIII in January 1547, returned to France with his family. With him he took his charts, maps and precious store of knowledge about the coastline of southern England. Although he had worked for the English king, it was the French crown who were the ultimate beneficiary of his labours.

nyfant Jsle

four Roche

porsail

obrenerac

Bretaigne

rouscou · le J. de bas ·
S. pol de leon · · Roche douve
morlays ·
port blanc · fept Jsles
entriguer ·
brehac ·

S. brieu

Guyene

dman · frelle ·

S. mallo ·

R. quy separe normandie s bretaigne
mont · s · michel · chauzer
potoxson · · granvile
granvile

66
NORD
constances ·
COSTAN
tm · balongnes
bayeux · cherbourg
quareten ·
BESSIM
H. de vire ·
estrehan ·

Jarzey ·
nez de car
teret
caxtet ·
mid · a Laigle ·
orom ·
S. germam · porbail · c. de hague ·
le
J. pele ·
hougue ·
barfleu ·
J.les · S · marcol ·

Guernzey ·

Long fheapped ·
monshole
monsbay · penfant ·
Lezart · helston
per m
falmouth ·
tregu
dodman ·
S. austel ·
fandu ·
Laftard ·
S. germans ·
plemouth ·
plinton · anstok ·

Saudestenr ·
ocanton ·
daxtmouth · totnes ·
afclipton · thiknel ·
nelfton ·
exmouth · exerz ·
califorde · stokland ·
homton ·
chard ·
lim · crokhon ·
brytport ·
fhirbon
weymowth · dorchester ·
wenrafter
roufhd fhaftsbery
clond
ford
corfe alias ·
the Jsle of portland
pole ·
fordingbru
hinton · milton · ringvode
zumsay
limington
bewlay · hampton
wiche
nedles
Wicht · yarzer
feche felde · porchester
fhelene
thornay · portsmouth · somexton
halsig · exton
haxant
waltin
vin fte

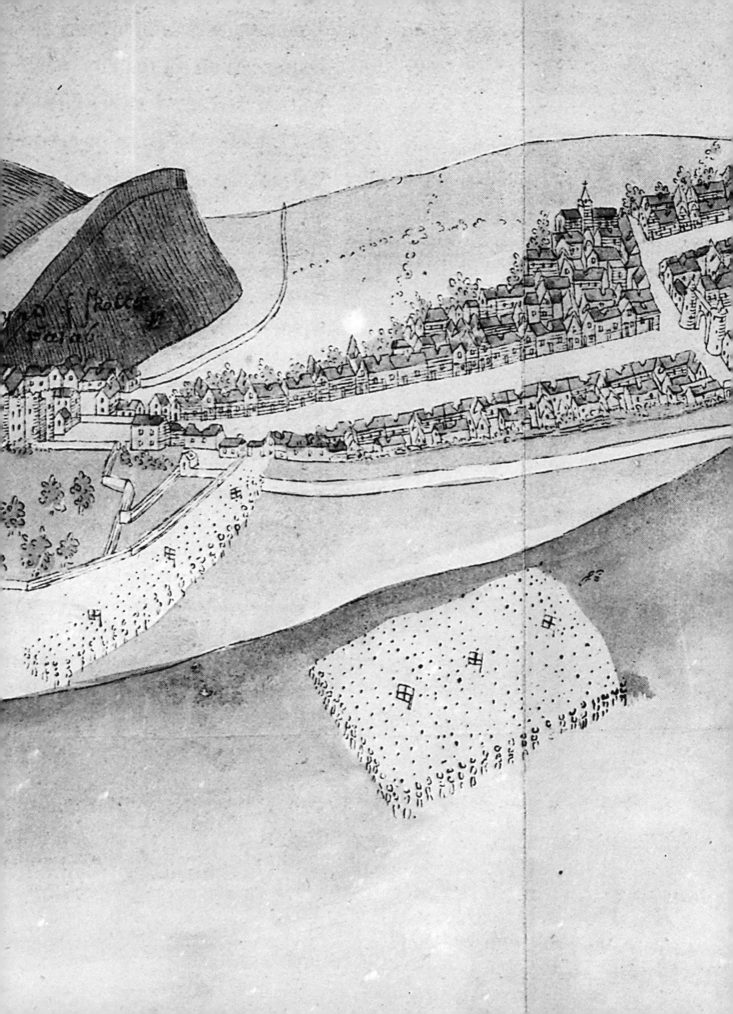

PLAN OF THE SIEGE OF EDINBURGH

The undulating landscape around Edinburgh is sketched out with precision and the detail of the streets between Edinburgh Castle and the royal palace at Holyroodhouse is intense in this 1544 map of Edinburgh. Only the presence of several units of troops, their squares bearing miniature English flags, reveal that this is no peaceful plan, but a depiction of the siege of the city by an English army under the Earl of Hertford in 1544.

The origins of the invasion of Scotland launched by Henry VIII were both immediate – in his spurned demand that the infant Mary, Queen of Scots marry his son and heir Edward – and deep-rooted, arising from the centuries-old tangle of rivalries between England, Scotland and France. Scottish relations with the new Tudor rulers of England after 1485 had got off to an uncertain start after James IV supported Perkin Warbeck, an impostor who claimed to be the Duke of York (and thus a rival claimant to Henry VII's throne). But by 1501, pragmatism had trumped the desire for revenge and Henry agreed that James could marry his daughter Margaret Tudor, a match sealed in 1502 by a Treaty of Perpetual Peace between England and Scotland.

Perpetuity in this case turned out to mean ten years and in 1513 James, fearing that France, a staunch bulwark against English ambitions, might buckle in the face of the Holy League between Venice, Spain, the Papacy and England, reneged on his treaty with Henry and marched south across the Tweed. At Flodden Field he paid the price of his broken trust: the English foot overwhelmed the Scots and James was killed, along with the flower of the Scottish nobility.

Scotland was left almost defenceless, and further weakened by feuding between the Duke of Albany, the regent for the infant James V, who headed the pro-French party, and Margaret, James IV's widow and the leader of the English party, who rapidly remarried Archibald Douglas, the Earl of Angus. The Earl soon came to dominate the regency. Only when James V gave Angus the slip in 1528 and drove him across the border into England, did Scotland begin to recover a shadow of independent will.

James had to rebuild, and in part he did this by strategic marriages, first to Madeleine, the daughter of Francis I of France, who unfortunately died six months later and then in 1538 to Mary of Guise, who came with a handy dowry of 150,000 livres. In 1542, however, James was offered the sovereignty of Ireland by a group of Irish chieftains keen to diminish English influence there. Henry used this affront as an excuse to invade Scotland, and a Scottish counter-thrust was roundly defeated at Solway Moss on 24 November 1542. Already ill, a broken-hearted James died, just after receiving news that Mary had given birth to a daughter, which left no male heir to succeed him.

The week-old queen became the subject of a tussle for her hand in marriage. Henry thought he had secured an agreement with the regent, the Earl of Arran, that little Mary should marry his son Edward. When the Scottish parliament revoked this promise, the outraged Henry declared war in December 1543, sending an army north from Newcastle, and an accompanying fleet sailing towards Scotland's east coast from Tynemouth. The army, some 10,000 strong included 6,000 border horsemen and 1,000 Yorkshire archers. Its commander, the Earl of Hertford, was ordered by Henry to 'put all to the fire and sword, burn Edinburgh, so razed and defaced when you have sacked and gotten of it what you can, as their may remain forever a perpetual memory of the vengeance of God . . . for their falsehood and disloyalty.'

Hertford carried out this 'Rough Wooing' with determination. Elaborate plans for the invasion had been pared back on grounds of cost to a simple raid on Edinburgh. On 2 May the English fleet was sighted off Leith and the next day they began to land. Light Scottish resistance could not prevent the English taking Leith and, having left a garrison of around 1,500 men, they marched on Edinburgh. There they were met by the town Provost Adam Otterburn, who offered to surrender it to Hertford on several conditions. But the Earl, though he had authority to burn Edinburgh, had none to negotiate its sparing, and refused.

English cannon blasted the town's Netherbow gate and the infantry surged through, killing hundreds of defenders and set to an orgy of destruction. Their enthusiastic looting, however, could not conceal the fact that their numbers, and even their heavy artillery, were no match for the fortifications of the Castle, defended by Captain James Hamilton, from where volleys of fire down the Royal Mile were taking a toll on the attackers. After a decent interval, Hertford ordered a retreat. On 14 May the harbour at Leith was burnt and, having knighted fifty-eight of his company for bravery, he sailed off, carrying £10,000-worth of looted goods and the prizes of the Scottish royal ships *Salamander* and *Unicorn* which had been berthed at Leith. Little had been achieved and Richard Lee, the royal engineer and former Surveyor of Works at Calais who drafted the plan of the siege, must have reflected that his plan showed little more than the virtuosity of Scottish military engineering.

see more on next page >

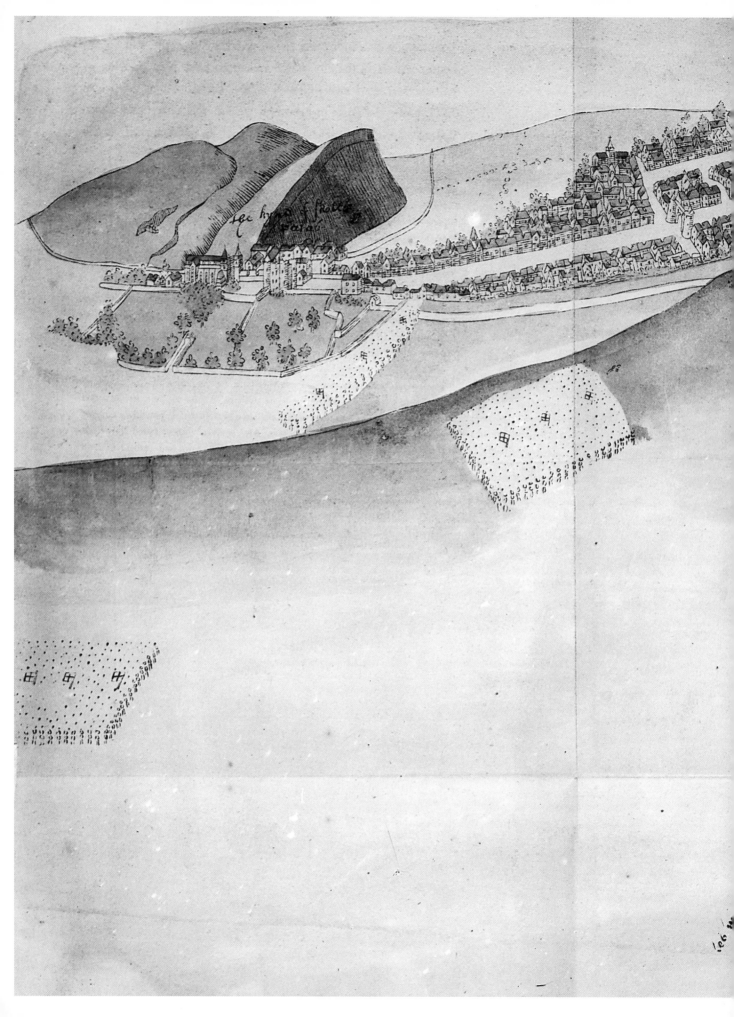

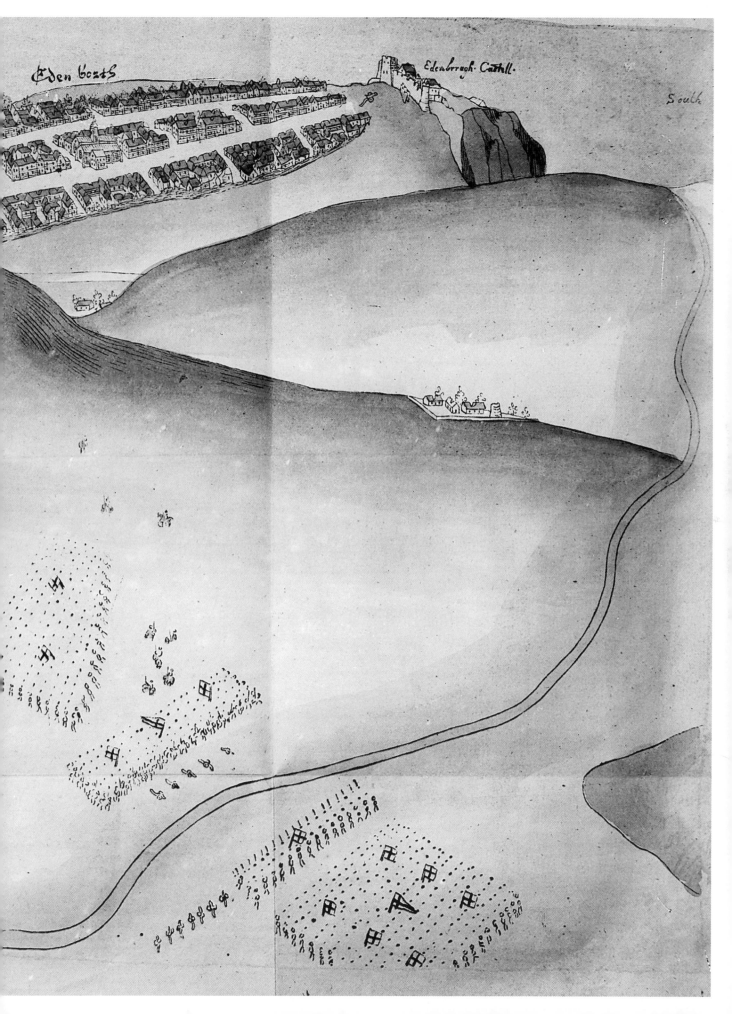

Eden boats

Edenbrugh. Castell.

South

The englishe horsemen.

The schotteshe horsemen rynnynge awaye.

The schotteshe battell.

The shurche.

PLAN OF THE BATTLE OF MUSSELBURGH

The sixteenth-century woodcut of the Battle of Musselburgh evokes all the drama of the battlefield at the dawn of the age of gunpowder. Massed ranks of Scots pikemen face off against the English divisions, as arquebusiers, artillery and ship-borne cannon blast away at the footmen in the last pitched battle fought between the armies of England and Scotland.

The Battle of Musselburgh (or Pinkie Cleugh) was fought on 10 September 1547, as part of the 'Rough Wooing', the attempt by the English to bind Scotland into an unfavourable marriage alliance between the infant Queen Mary and Henry VIII's young son and heir, Edward. The death of Henry in January 1547 did nothing to dampen the ardour of his advisers to pursue the match, well aware that, if successful, it would amount to a virtual annexation of Scotland. Needless to say the Scottish regent, the Earl of Arran, was equally as opposed to the young couple's betrothal and he watched with anxiety as news of the mustering of an English army reached Scotland through his agents in London.

By early September, the Duke of Somerset, the Lord Protector (and uncle of the new king, Edward VI) had crossed into Scotland, accompanied by an army of some 18,000 men, including 6,000 cavalry (many armed with arquebuses), a few hundred German mercenaries and a large contingent of county levies equipped with longbows and bills. Somerset pushed along the east coast to ensure contact with the English fleet, while Arran was kept off balance by a diversionary attack into Annandale in the west, which ensured he could not be quite certain of the direction of the main English attack.

By 9 September, Somerset's force had reached Musselburgh, east of the river Esk, where he found the Scottish army, some 22,000-strong, barring his way along the coast to Edinburgh, 10 km (6 miles) away. The Scots were mainly pikemen, with a leavening of light cavalry and were outgunned (if not outnumbered) by the English. As if to compensate for these shortcomings, Arran issued a chivalric challenge, sending forward a herald and trumpeter to challenge 2,000 of the English cavalrymen to a skirmish with an equivalent number of Scots. Somerset accepted, and his contingent under Lord Grey proceeded to cut to pieces the Earl of Hume's mounted squadron.

Arran tried a further challenge to single-combat with Somerset, but the Lord Protector, impatient with his equivocations, declined to pick up the gauntlet. Aware that his ranks might be shredded by the superior English artillery, the Scottish

see more on next page >

commander then ordered his men forward across a Roman bridge over the Esk, hoping to force a close-quarters fight that would nullify Somerset's advantage. Although Arran's right flank managed to push back the English cavalry and make some progress, his left was bombarded by the English ships and he soon found himself under fire from three directions. Scottish morale wavered, buckled and then broke. Thousands died as they fled from English gunfire, bogged down in the marshy ground or drowned as they tried in desperation to swim the swift waters of the Esk.

Despite the gravity of the Scottish losses – perhaps 6,000 men died – Somerset failed to capitalize on his advantage. He marched only as far as Leith before, with supplies running low – he turned tail and marched back to England leaving a series of garrisons including at Roxburgh and Broughty Craig on the Tay.

The Scots soon raised another army and drove out most of the English garrisons, while a French expeditionary force which landed in June 1548 helped deter any further English raids. In 1550 peace was signed with France at Boulogne, and the next year the Treaty of Norham confirmed the English abandonment of its positions in Scotland. The main effect of Musselburgh was that Mary, aged five, was smuggled out of Scotland sent to France in 1548, where she was betrothed to the four-year-old dauphin Francis.

In Mary's absence, her mother Mary of Guise tried to stave off the advances of the Reformation in Scotland. It seemed as though her success was guaranteed when Edward VI died in 1553 and his Catholic sister Mary ascended to the English throne, but in the end it was not to be. Mary, Queen of Scots married Francis in 1558, and after the death of Mary Tudor later that year, she was regarded by Catholics as her legitimate successor. Francis, who became king of France after the death of his father, Henry II in 1559, pushed his case too far, quartering his arms with those of England as well as Scotland, implying his intent to depose his wife's Protestant cousin Elizabeth. This guaranteed renewed antagonism between Scotland and England. However, Francis died in 1560 and the following year Mary returned to Scotland, stripped of French support.

Although there were to be no more invasions, and no more battles such as Musselburgh, when Mary, who had become increasingly unpopular in a Protestant Scotland, fled south to England in 1578, her cousin Elizabeth refused to see her and had her imprisoned (and, ultimately, executed). The 'Rough Wooing' between England and Scotland had failed, although, ironically it was a child of Mary, Queen of Scots who united the two crowns when her son James VI succeeded Elizabeth in 1601, but not in a manner that anyone in 1547 could ever have expected.

THE.ENGLISHE.VICTORE.A
MVSKELBR

The cartes.
The hill of mvoeſelbrughe.
The engliſhe battell
The engliſhe horſemen.
The contre lothiana.
The ſchotteſhe battell
The engliſhe lyghte horſemen.
The engliſhe campe.
The ſhu
Mvorſelbrughe.
The engliſhe ſheppe.

YNSTE.THE.SCHOTTES.BY.
HE. 1547.

The hill.

The grene hill.

The schotteshe
horsemen
rynnynge
awaye.

The abbe of Edenburge.

Edenburge.

The grene hill.

The caster

The englishe campe.

The schottesche campe.

Litte.

The englishe galls.

THE.   SCHOTTES   CHE   SEE.

MAP OF THE DEBATABLE LANDS

To the north lies Scotland, to the south England and in between, on this 1552 map of the 'Debatable Lands', a small area that lay in dispute between the two countries and where the uncertainty over ownership had allowed a culture of banditry to flourish. The lines on the map mark the competing proposals between the English and Scottish commissioners who met to resolve the problem and the final border that was accepted in order to stamp out this haven of lawlessness.

The border between the two countries had long been fluid, and it was for many centuries unclear where it would settle. In the seventh century, the English kingdom of Northumbria had held sway as far north as Edinburgh until a crushing defeat inflicted by the Picts on King Ecgfrith at Nechtansmere in 685 pushed central Scotland out of the English sphere. In the twelfth century it seemed as though the northern counties of England would be absorbed by Scotland, when David I intervened in support of his niece Matilda, who claimed the English crown after the death of Henry I. In 1136, he marched into England and occupied a string of castles, including Carlisle, Alnwick and Newcastle. Land-hunger overcame family loyalty, however, and he made peace with Matilda's foe, King Stephen, in exchange for keeping Carlisle (which, despite reneging and then suffering a stinging defeat at the Battle of the Standard two years later, the Scots retained until 1154).

Then, ever eager make capital out of English dynastic squabbles, Alexander II invaded England in 1215 in support of Louis of France's attempt to overthrow King John (having been invited in by baronial malcontents), leading his army all the way to Dover. The death of John, paradoxically, thwarted him, for the English barons then united behind his young successor, Henry III. When Louis came to terms with the new king, Alexander was left isolated and had to evacuate the northern English counties which he had occupied. In 1237 he negotiated the Treaty of York with Henry, which set the border much along its modern course, with the exception of a 16 km (10 mile) by 6.5 km (4 mile) tract in the west between the rivers Sark and Esk.

Although relations between England veered in the following centuries between poor and downright hostile, with the invasions launched by Edward I of England from 1296 marking but a low-point, exactly what constituted England and what made up Scotland was largely not in dispute. Except, that is, in the case of those 'Debatable Lands' in the west. Here, on both sides of the permeable border, families of 'border reivers' flourished, raiding at will on the other side and occasionally providing martial support to their theoretical sovereigns, as when a large contingent of Scottish reivers fought at Flodden in 1513.

Largely, though, the reivers were the source of intense irritation to the authorities who resented the depredations of families such as the Armstrongs on the Scottish side and the Grahams on the English. In 1530 James V of Scotland took decisive action and had Johnnie Armstrong of Gilnockie and thirty-one leading reivers hanged, while in 1531 the English and Scottish authorities declared that 'All Englishmen and Scottishmen, after this proclamation made, are and shall be free to rob, burn, spoil, slay, murder and destroy all and every such persons, their bodies, buildings, goods and cattle as do remain or shall inhabit upon any part of the said Debatable Land without any redress to be made for the same.'

Finally, in 1552 a commission was established to settle the border once and for all, and, it was hoped, to reduce the Border reivers' freedom of action. Sadly, the raiding continued and it took a renewed effort under James VI of Scotland when he ascended to the English throne (as James I) in 1603 to impose relative peace on what he now called the 'Middle Shires'. New courts were established to try offenders and, given that reivers could not now evade justice for offences on one side of the border by fleeing to the other, the level of their activities subsided. If not exactly peaceful, the borderlands were no longer debatable.

see more on next page >

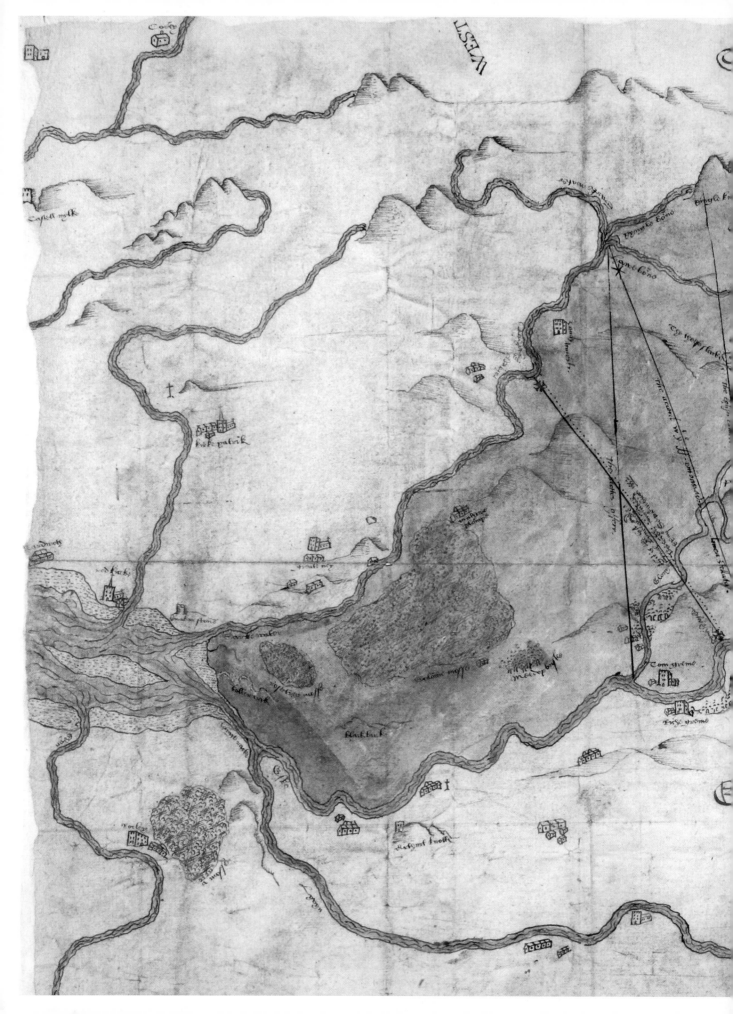

CHEAM PARK AND NONSUCH

The colourfully drawn patchwork of fields, meadows, copses, village houses and churches form, on the face of it, the backdrop to a court case between the tenants and farmers of the royal estate at Malden and those of the villages of East and West Cheam over the rights to grazing on a patch of common ground. Yet, situated at the top left of this 1553 map lies the outline of Henry VIII's grand palace of Nonsuch, a project that gave architectural form to the king's vanity and which became a microcosm of the politics of the age.

By 1538, Henry had suffered a brutal few years. He had in rapid succession divorced one wife (Catherine of Aragon), beheaded another (Anne Boleyn) and seen a third (Jane Seymour) die in childbirth in October 1537. Almost precisely a year earlier the Pilgrimage of Grace, a religiously inspired uprising in Yorkshire and Lincolnshire, had come perilously close to unseating him from the throne, as devotees wearing the badge of the bleeding heart of Christ had gathered under the leadership of a charismatic lawyer Robert Aske. Before long 40,000 men were following his banners, demanding the removal of 'heretical' bishops such as Cranmer and the restoration of the old Catholic faith. Only the false promise of a pardon delivered by the respected Duke of Norfolk diffused the momentum of the revolt and allowed Henry's men to arrest Aske and hang over 200 of the rebels, stamping out the spark of dissent in the north.

Henry was ageing, corpulent and suffered persistent ulcers on his leg; a far cry from his own self-image as a chivalric paladin bestriding the battlefields of Europe. His twenty-year rivalry with Francis I of France seemed to have been resolved in favour of his rival, who had begun building a lavish chateau at Chambord in 1519. Determined not to be outdone, Henry threw himself into a frenzied campaign of building, most notably in 1531–6 at Hampton Court (the former home of the disgraced Cardinal Wolsey).

In 1538, Henry turned to another project. He needed a hunting lodge close to London to which he could travel without a long journey on horseback, which his ill-health now made a tiring prospect. His eyes soon lit on the manor owned by the Codyngton family near Cheam in Surrey. The Codyngtons were made a peremptory offer for their lands, and building began almost immediately. The church of Cuddington was demolished to make way for the new palace's inner court, whose splendour soon became so apparent, it was named Nonsuch – a way of claiming that it had no equal in Europe.

No expense was spared in Nonsuch's construction. The total cost has been estimated at around £24,000 (around £14 million at today's prices). By August 1538, the workforce there included 120 carpenters, 14 carters, 11 chalk-diggers and 2 thatchers. A thousand deer were gathered from other royal hunting parks to populate the Great Park to be laid out around Nonsuch and it took 600 cart-loads to convey the fencing necessary to keep them from straying.

Curiously, once the palace was complete, Henry rarely visited it, doing so only four times, the last in December 1546 a month before his death. Although his daughter Mary sold it to the Earl of Arundel, it came back under royal control during Elizabeth I's reign, and she spent part of many summers there. It was the site of a conference in 1585 which agreed a treaty with the Dutch – the first international treaty of the Netherlands – by which the English agreed to provide 7,000 troops to help the fledgling republic in its rebellion against Spain. It was also at Nonsuch that, on 28 September 1599, Robert Devereux, Earl of Essex, and the former favourite of the queen, caused a scandal by bursting in on her unannounced in her bedchamber. He had dashed from Ireland, where he had negotiated a truce with the Irish rebel leader, the Earl of Tyrone, rather than defeating him, as had been Elizabeth's command. Muddied, breathless and impertinent, Essex paid the price for his presumption and was detained, tried – on rather more substantial suspicions of treason – and executed in February 1601.

Nonsuch's subsequent history was more tranquil if short. The end came for it in 1682 when the palace was demolished on the orders of Barbara Villiers, Countess Castlemaine, the mistress of Charles II, to whom the king had gifted it. An inveterate gambler, Lady Castlemaine ran up staggering debts (once reputedly losing £25,000 in a single night at cards) and could not afford to keep the palace up. A monument to the folly of one king's attention to his mistresses, it was brought low by the mistress of another.

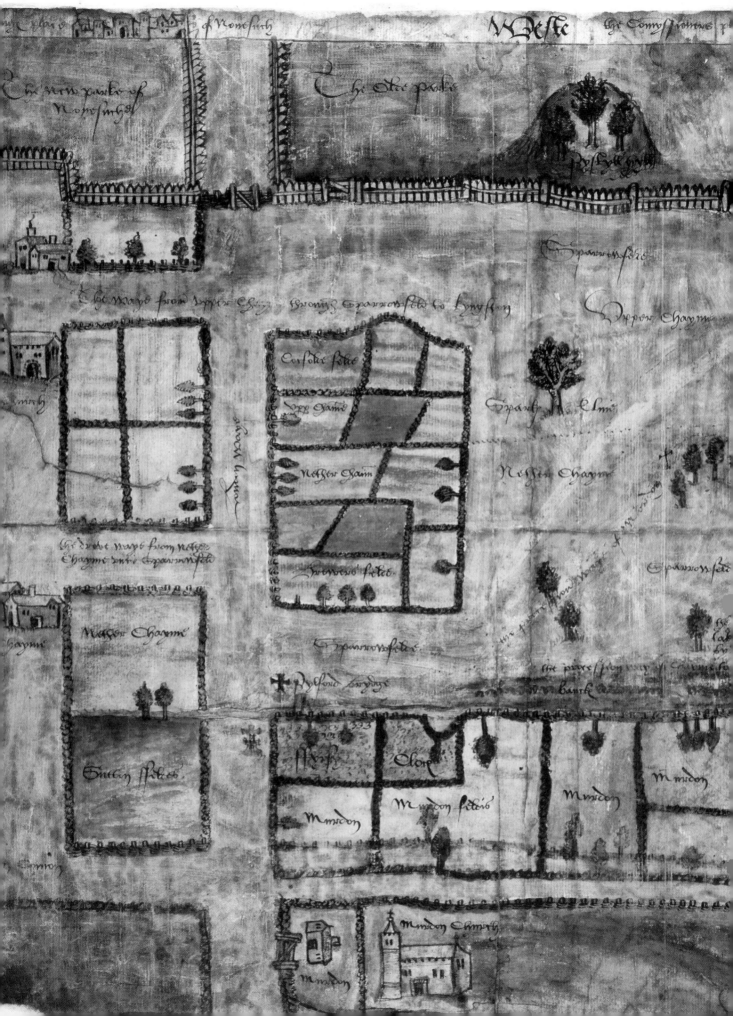

COPPERPLATE MAP OF LONDON

It is the earliest surviving true map (as opposed to panorama) of London. This 1558 copperplate engraving formed the basis of plans of London for the next half century. It shows a city (etched in reverse), whose 100,000 people were packed into an area of only around a couple of square kilometres (1 square mile), with teeming alleys and labyrinthine lanes only occasionally punctuated by the spire of one of London's many churches. London Bridge can be seen at the bottom of the plate.

The plates (of which only three of the original fifteen survive, while no printed version of the map now exists) were probably made late in the reign of Mary Tudor by German merchants who were lobbying her husband, Philip II of Spain, to maintain their privileges in London. Its date is indicated by the absence of the Royal Exchange (only opened in 1571) and the siting of a cross in the churchyard of St Botolph-without-Bishopsgate which was destroyed in an anti-Catholic riot a few months after the accession of Elizabeth I in 1558.

Marriage and Catholicism – these were the two themes which dominated the life of Mary, Henry VIII's eldest surviving child. Her birth in 1516 had been a bitter disappointment to the king after the death of two sons who had only survived a few weeks. Although she was cherished, and received an excellent education (learning to speak and read French, Spanish and Latin), the lack of a male heir bothered Henry, and Mary became the subject of protracted negotiations in which her hand in marriage was variously promised to the French dauphin, Emperor Charles V and Duke Philip of Bavaria.

As a result of the annulment of Henry's marriage to her mother, Catherine of Aragon, Mary was declared illegitimate in 1533 and lived a shadow-life in exile from the court. Only in 1537, when the king's third wife gave birth to a son, the future Edward VI, thus rendering both Mary and her younger sister Elizabeth irrelevant to the dynastic succession, was she welcomed back into the royal fold. When the old king finally died in 1547, Mary was forced into a delicate balancing act, keeping true to her Catholic faith while trying to distance herself – but not too obviously – from the increasingly radical Protestantism of Edward VI's advisers. The outbreak of Kett's revolt in Norfolk in 1549 – in fact in reaction to land enclosures in East Anglia – posed a particular danger to her, as her enemies could have used it to accuse her of plotting against the king.

Mary survived through this crisis, and another in July 1553 when the death of her sickly brother was followed by an attempt by the Duke of Northumberland to install Lady Jane Grey (daughter of the Duke of Suffolk) on the throne in order to ensure a Protestant succession. Despite the odds against her, public sentiment saved Mary and revolts in Norfolk and Kent, and the defection of the fleet to her cause led to the collapse of Northumberland's plot.

As queen, Mary moved rapidly to sweep away the apparatus of the Protestant Reformation. Many of the Protestant leaders were arrested and charged with high treason, married priests were deprived of their benefices and Protestant refugees from France and the Netherlands were arrested. She also opened negotiations for a marriage with Europe's leading Catholic monarch, Philip II of Spain, which took place in Winchester on 25 July 1554.

Mary's determination to impose Catholicism lost her much of the popularity which had accompanied her accessions. Wyatt's Revolt, an uprising in strongly Protestant Kent in January 1554 had already served as a warning, but Mary pressed ahead. In November 1554, the Act of Supremacy, by which Henry VIII had installed himself as head of the English church, was repealed and in early 1555 the burning of leading heretics (as recalcitrant Protestants were now viewed) began. Chief among them that year were Nicolas Ridley, bishop of London, and Hugh Latimer, bishop of Worcester, and, in March 1556, Thomas Cranmer, the archbishop of Canterbury and the principal architect of Henry's religious settlement.

Mary's reign ended in disillusion. Public opinion had turned against her and her memory would be damned by the lurid accounts of the burnings published in George Foxe's *Book of Martyrs* published shortly after her death. The marriage with Philip II yielded no heirs and led to England becoming embroiled in a war with France and the resultant loss of Calais, the last English outpost on French soil, in January 1558. It was said, unkindly, that when she died, an isolated, embittered woman on 17 November that year, that 'Calais' would be found engraved upon her heart.

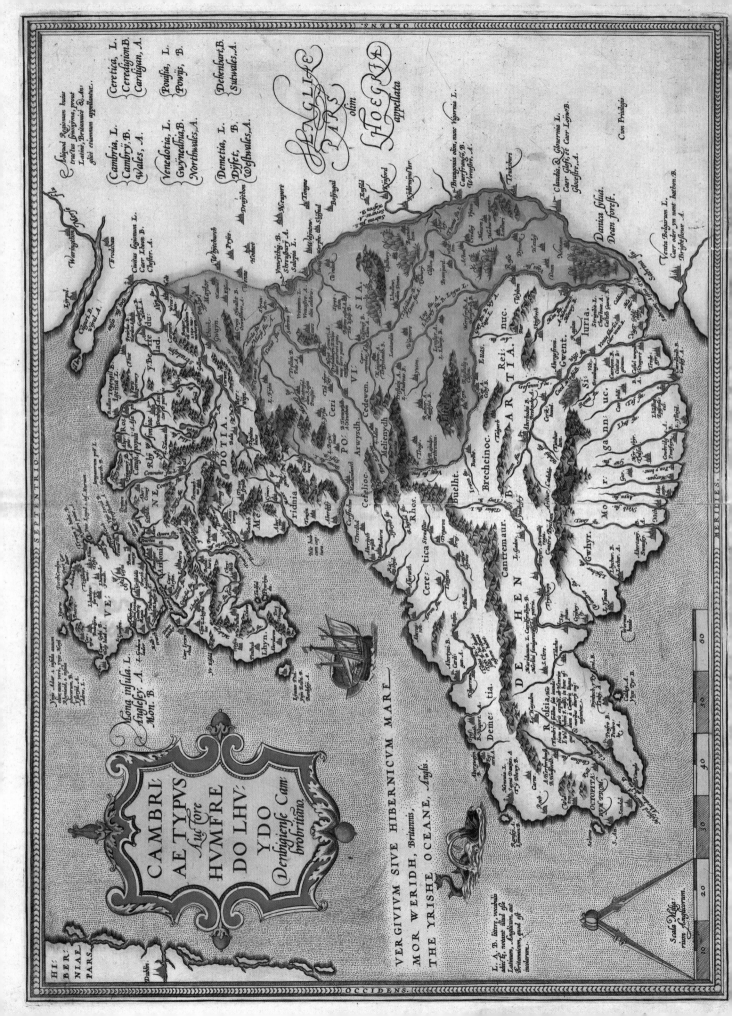

CAMBRIAE TYPUS,
Humphrey Llwyd

It is a map which looks back to Wales's distant past and forward to its future as part of a union with England (and ultimately Scotland and Ireland). Humphrey Llwyd's 1573 *Cambriae Typus* is the first printed map to portray Wales as a discrete entity – in some ways paradoxically, as it was just thirty-seven years since the embers of its existence as an independent kingdom had been extinguished.

Llwyd was a physician and a member of the well-to-do Welsh gentry class that had accommodated itself to English rule and prospered as a result. Educated, like many ambitious Welshmen, at Oxford University – for Wales had no university of its own – he became interested in cartography, and after an introduction to the eminent Dutch mapmaker Abraham Ortelius, he contributed the Welsh section which appeared in a supplement to the Dutchman's world atlas, the *Theatrum Orbis Terrarum*, in 1573.

Llwyd's map is reasonably accurate, though the Welsh coastline wavers rather uncertainly from its true course and the country's boundaries extend to the River Severn, encompassing areas that were really English. It is, however, unashamedly antiquarian, including place-names, such as Demetia for Dyfed, and Venedotia for Gwynedd, which owe more to the great second-century AD geographer Ptolemy of Alexandria than to sixteenth-century reality. Yet perhaps skipping back to the ancient past was a way of ignoring the Middle Ages, which brought both triumph and agony to the Welsh and ended with the real prospect of the loss of their identity.

The medieval kingdoms of Wales had first collided with England in the aftermath of the Norman conquest, when William the Conqueror led an army as far as St David's in 1081. The English advance thereafter was largely a freelance affair, led by magnates such as Richard Fitz Osbern, Roger of Montgomery and Hugh of Avranches who set up great marcher lordships which absorbed much of the eastern borderlands and the south of Wales. A general uprising in 1094 prevented the total disintegration of Wales, but endemic squabbling between Gwynedd, Dyfed, Powys and the other Welsh principalities impeded the development of a single united front against the English. Cultural influences from outside, such as the importation of continental monasticism – there were fifteen Cistercian houses in Wales by the late twelfth century – also softened resistance to the outsiders.

Llywelyn Fawr ('the Great') of Gwynedd finally established the supremacy of his principality over the others, but his death in 1240 led to a civil war that further sapped the Welsh capacity to resist English aggression. Finally, his grandson Llywelyn ap Gruffydd was able to take advantage of England's weakness during the struggle between Henry III and his barons under Simon de Montfort, to extract recognition as Prince of Wales in 1267. After Llywelyn refused to do homage to Edward I, a new war broke out, which ended with his assassination in 1282, the studding of north Wales with a series of near impregnable castles such as Caernarvon and Conwy and, in 1284, the passing of the Statue of Rhuddlan, which withdrew all the privileges Llywelyn had gained and made clear Wales's subject status.

Discontent festered in Wales, but no serious revolt threatened English rule until that of the Welsh nobleman Owain Glyndwr in 1400. An initial campaign of hit-and-run attacks nettled the English, but the capture of Conwy in 1401 rallied Welsh opinion behind the rebels, as one-by-one previously unassailable strongpoints such as Harlech and Aberystwyth fell. Despite the symbolic holding of a Welsh parliament in Machynlleth in 1404, it was all an illusion and when an Anglo-French truce in 1407 ended all hope of military support from France, Henry IV was finally able to concentrate on Wales, and the revolt was put down.

Welsh hopes were raised by the succession of one of their own, Henry Tudor, as king of England in 1485. Yet his priorities and those of his son, Henry VIII, proved to be the overall stability of the realm and that meant ensuring that Welsh exceptionalism never produced another Glyndwr. In 1536 the Act of Union was passed, which virtually extinguished Wales as a separate legal entity; the marcher lordships were swept away and the country divided into thirteen shires. Most Welsh nobles and gentry accepted the new regime and ambitions henceforth focused on rising in the English court hierarchy, rather than uprisings at home. Only the continued strength of the Welsh language (around 90 per cent of the population were still monoglot Welsh speakers) preserved the spark of Welsh cultural identity, aided by the publication of the first complete Welsh Bible in 1588.

It was in this atmosphere that Llwyd produced his map, which was very much a portrayal of a once and future kingdom.

CAMBRIDGE TOWN PLAN

The Cambridge colleges cluster around the town centre and the Backs (along the banks of the River Cam) in this 1574 plan, dominated by the outline of King's College Chapel. Although in one sense little has changed in the succeeding four centuries – save the giveaway Tudor garb of the fisherman opposite Clare College – in another, the nature of early modern universities was very different to those of the twenty-first century.

Throughout the early Middle Ages, education remained firmly the domain of monasteries and then, slightly later, cathedral schools. But these institutions could not provide sufficient trained men to staff the growing royal or local administrations which characterized countries such as Italy, France and England. To fill this gap, a class of secular school, the *studium generale* or university grew up, beginning with Bologna in 1088 and Paris soon after.

Oxford was the first English university, operating in some form by 1117, but more clearly established by 1167. It had no rival until 1209 when Cambridge was founded by academics fleeing unrest at Oxford (after which two scholars were executed). They then had no English rival until the nineteenth century when University College London received its charter in 1826. This duopoly was threatened only a blip in 1261–5, when an ephemeral university was functioning in Northampton, founded by students fleeing riots in Cambridge; university life was clearly a rumbustious affair.

Scotland received its first university, St Andrews, in 1413 followed by Glasgow in 1451, Aberdeen in 1495 and Edinburgh in 1583, meaning that it had, for the following 250 years twice the number of universities as England, despite its far lower population. Ireland had a single university, at Dublin, founded in 1592. Wales had none at all until the foundation of St David's College, Lampeter in 1822, having to content itself with Jesus College, Oxford, which was established in 1571 to cater for the education of Welsh students.

Medieval universities were, at their core, intended to train students to take up role in the law, medicine, teaching or the Church. Their focus was on the ancient classical curriculum of the *trivium* (grammar, logic and rhetoric) and the 'advanced' studies of the *quadrivium* (music, arithmetic, geometry and astronomy). Students achieved this by careful digestion of a preordained corpus of knowledge, attending lectures on the core curriculum for some years before being trusted to attend disputations, or debates, given by university masters.

The distinction between undergraduates and graduates only emerged in the early fourteenth century when younger students began to be admitted (the statues of what is now King's College, Cambridge ordained that they must be at least fourteen years old). At about the same time the colleges, which had originally been founded as centres for graduate students (beginning with Merton College, Oxford in 1264) began to transform themselves into something approaching today's communities of undergraduates, graduates and academic staff; whereas previously most undergraduates had lived in halls or hostels (of which there were sixty-nine at Oxford in 1440), by 1552 they almost all lived in the colleges.

University life could be taxing. The statutes laid down requirements for the attendance at lectures – at Oxford candidates for the master's degree had to carry out daily disputations in logic during Lent, save on Fridays which was, curiously, reserved for grammar. The day began at 5 a.m. and in the sixteenth century lectures at Corpus Christi College, Cambridge started at 6 a.m. (with a bracing discourse on Aristotle's philosophy) and carried on until late afternoon, punctuated by prayers and expositions of passages of scripture. Lest revelry get in the way of study, most forms of entertainment were forbidden, including all forms of gambling, dancing, and even the playing of chess, while the keeping of dogs or hawks was also strictly prohibited.

The number attending university was very small; in 1400 there were only about 2,000 students in England (three-quarters at Oxford and 500 at Cambridge), which rose to 3,000 by 1450 (when Cambridge had grown to 1,300 students) and reached a high-point of 6,000 in the 1630s (when the population of England was around 4.5 million). Yet, though the student body was a highly selective one, the tradition of public service that grew up during Tudor times meant that university education provided a clear avenue of advancement for the gentry and minor aristocracy. Men such as Thomas Cranmer, who helped guide Henry VIII towards the establishment of the Church of England, studied law at Jesus College, Cambridge and John Pym, whose masterful opposition to Charles I in the Long Parliament helped provoke the English Civil War, studied at a forerunner of Pembroke College, Oxford. They showed that the body of educated men who issued from the first British universities had the power to shape and to challenge the existing order in a way none of their class could possibly have done before.

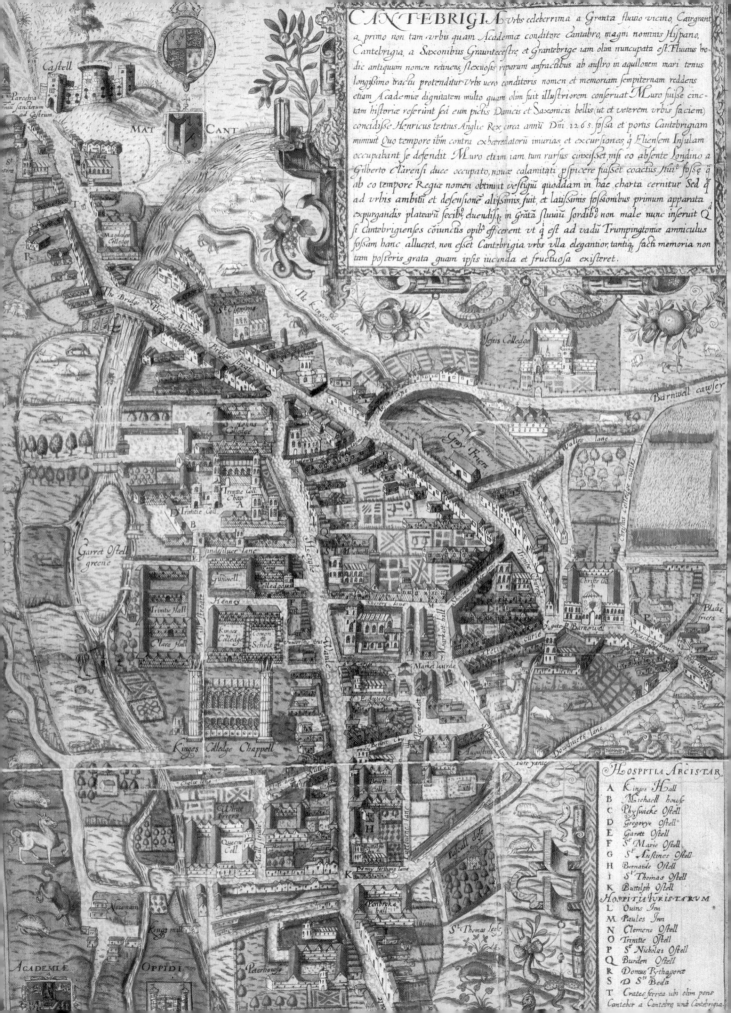

Pentire point

Ferrabery

Tintagell cast

BOSCASTLE
Sct gennyns
Iacobsto
Pouns

TREVENA
Treualgay Minster
Tintagelle ne
Lesnethe
Sct Iulyet
Warpestow

LESNOWYTH
HV
Dauidstow
Otterhm
Treneglos
Tremer
Pen

The black rock
Padstow hauen
The Gull rock

Sct Teathe
Lanteglosse
CAMLEFORDE
Sct Aduen
Enyam flu
Laneast
Trewe

Sct meren
PADSTOW
Sct Endelion
Sct Mynner
Sct Kew
Michelstow
Tresmure
Sct Clether
Lawannok
Trelaske

Sct Euall
Sct Petheryk
Sct Eruan
Sct Tissye
Tredenik
Sct Breake
Waddbridge
Iglessale
Sct Tudye
Sct Bruard
Rowtor hill
Brounwellye hill
Alternun

Lanherne
Sct Maugan
PYDER
HV
Treuerder
Helegen
Sct maben
TRIGGE
HV
Northhill

lum
ua
Sct
COLVMBE MAGNA
pua
The 9 Stones
Helan
Blisland
Temple
Lomera flu
EA

tryse
Cuswarthe
Sct Wen
Wethiell
Worlegan
Len

Sct Enoder
Sct Laurence
BODMAN
Cardenhm
WEST
Fowe flu

Melader
Vale flu
Penuenten
Sct Denis uale
Roche
Laneuet
Lanbetherek
Glyn
HV
Sct Neot
Sct Clare
Bickton
Sct Iue

ladok
Sct Steuens
Luxelion
Lestermar cast
Reprin
LISKERD
Quethiok

POWDER
HV
LESTETHIEL
Pinrok
Cortyther
Mynhenyet

GRAMPOVND
Me wan
Sct Blaies
Lanleueray
Pelyn
Pill
Byconek
Sct Kyngtons h
Brodok
Low flu
Gayne
Chafrench
Wotto
Lanrak

rede
aus tell
Sct Winnow
Tethe
The Fyrynghowse
Duloo
Trewargo
Moruale
Sct Germans

ND
Curwarder rock
Trewardrath
Golant
Lanreath
Trewardrath bay
Sct Vepe
Haule
Plenynt
Porpyghan
Sct Merten
Shewok
Tregovik

Sct Tue
Michaell
Guran
Baudregan
Pentuar
Menegesse
Chappelland
FOYE
Polruan
Sct Siuers
Lanteglos
Tallande
Lansalowes
Low
Sct Michaels Insul
Crost hole

Porteus ie
The black rock
The black heade point
Foye hauen
Blaksto l point
Knaueland point
Poulper
Tallande point
Sothan bay
Long Pone

Dudman point
The Carnasse

COUNTY MAP OF
CORNWALL, Christopher Saxton

Adorned with fish, sea-monsters, ships and surveyor's tools, Christopher Saxton's 1576 map of Cornwall gives a picturesque view of the county, yet it also formed part of the very first proper atlas of England and Wales, at a time when the Tudor government's concern for stability meant it more than ever needed accurate knowledge of the lands over which it ruled.

Born to a family of modest means in West Yorkshire, Saxton served an apprenticeship to the local vicar, John Rudd, who had been commissioned in 1561 to carry out a survey for a national map. Rudd's venture was never completed, but in 1574 Saxton was instructed by Thomas Seckford, a court official, to undertake the task instead. After four long years traversing the country, the result was an atlas of thirty-five colour maps, which was published in 1579 as the *Atlas of the Counties of England and Wales*.

The map of Cornwall, completed in 1576, portrayed – if not wholly accurately, as far as the coastline goes – an area whose distance from the court at London made it a particular concern for William Cecil, Lord Burghley, Elizabeth I's Secretary of State, who took a close interest in the cartographic progress of Saxton. Cornwall had long been at the periphery, both geographically and politically, and in perennial danger that neglect from London might breed resentment and revolt.

Cornish particularism stemmed from its having been the last area to succumb to the tides of Anglo-Saxon invasions that swept over England from the fifth century. It had retained its independence as a last Celtic bastion until 838, when King Egbert of Wessex defeated a joint Cornish–Viking host at Hingston Down (and perhaps even longer, as a Hywel 'king of the Cornish' gave his fealty to Athelstan in 927). Although William the Conqueror marched into Cornwall in 1067 and put down disturbances there, the county remained semidetached from England, its self-identity nurtured by the long resistance of the Cornish language to the encroachment of English.

In 1227, in a bid to bind Cornwall more closely to the crown, Henry III invested his brother Richard with the earldom of Cornwall and the county sent two knights and six representatives of the towns to most medieval parliaments. However, in Tudor times the country reasserted itself, reacting to the instability of the times, both political and religious, with a series of revolts. In 1497, led by a St Keverne blacksmith, Michael Angove, 10,000 Cornish insurgents marched all the way to Kent and, camped at Blackheath before being defeated, as few nobles rallied to their cause. Just a few months later, however, the pretender Perkin Warbeck landed in Cornwall, claiming to be Richard, Duke of York (whom in fact Richard III had had murdered in the Tower of London). Following a failed siege of Exeter, the rebels were cornered at Taunton and, after Warbeck fled, they surrendered and threw themselves on Henry VII's mercy (which came at the cost of £14,000 extracted for the issue of pardons).

Cornwall was religiously conservative and anger at the progress of the Reformation boiled over in 1548–9 when rebels assembled at Bodmin carrying the sign of the consecrated host under a canopy and demanding the restoration of Catholicism. They had the misfortune to face a particularly intransigent opposition, as the young Edward VI (who became king in 1547) was a staunch Protestant whose imposition of a new *Book of Common Prayer* in 1549 gave the Cornish rising the nickname of the 'Prayer Book Revolt'. The rebels took the traditional path of besieging Exeter, but were defeated at Clyst Heath and then fled back across the traditional Cornish boundary line of the River Tamar, after which the uprising collapsed.

Edward VI's imposition of radical Protestantism, which included the issue of the Forty-Two Articles, an uncompromising anti-Catholic manifesto, by Archbishop Cranmer in 1553, was overturned by his sister Mary, an equally ardent Catholic, after her accession later that year. Her sister and successor Elizabeth I needed to tread more carefully. Although she reasserted Protestantism in a more moderate guise (which formed the origins of the Anglican church), her failure to marry (and consequent lack of an heir), tense relations with her cousin Mary, Queen of Scots and the ever present threat of invasion from Spain meant she could never sit easy on her throne. A major attempt to invade was not launched by the Spanish until the Great Armada of 1588, but the fear that they might do so underlay royal sponsorship of project such as Saxton's great county maps of England and Wales.

see more on next page >

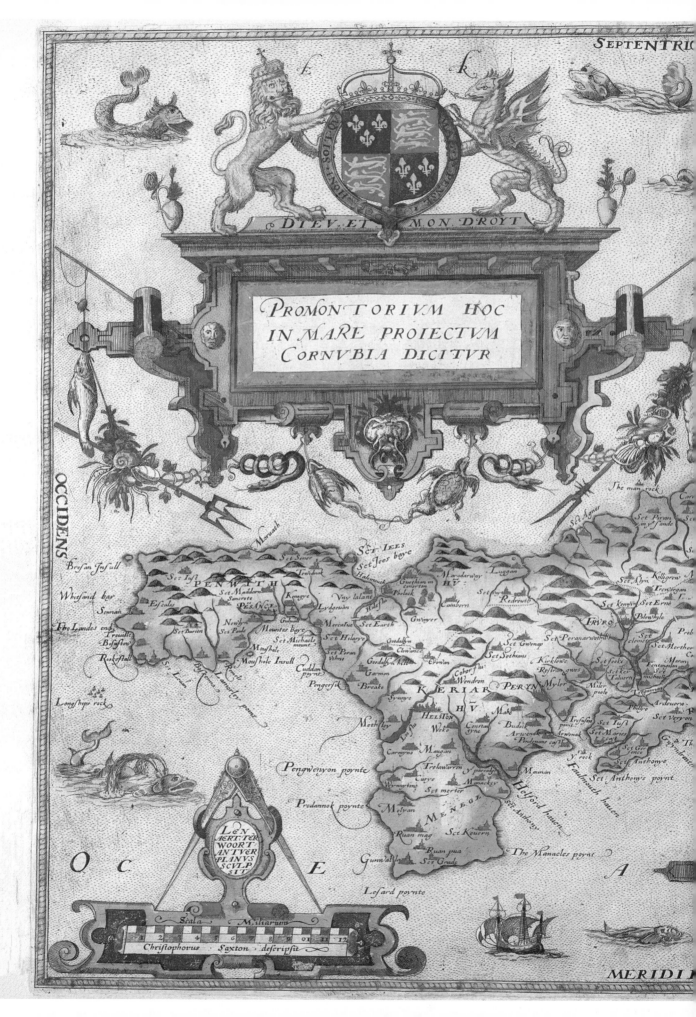

Factum est hoc
opus An° Dni 1576
et D. ELYZABETHE
Reginæ 18

Hartland
point
Hartland

HARTON
Welcombe   Clauelle
Morwynstow
Tunacombe
North lee
Stow   PAR
Kilkhmiton
Norton   Bradworthy
STRATTON   Sauldon
Launcels   Droxboro
Pancretwell
Whalesboro Northïa
churche   Bridgerewell Piworthye HOLSWORTHYE
Wolston   Moris
Pounstoke HV.
Whitstone   Clauton
Scé germyns Sct marre Wike Ygboro Tamerton
Iacobstow   Tecot   TE
Boyton   Luffencote
North Pit   OF
herwin
Sct Giles
Warpstow   Wirrington
Tremene
Trencolos Penhale   Sct Steuens LYFTON
Tresmure Egleskey   Ledesfa
Sct Clether Laneast LAVNSTON
Ewyns flu Trewen   Lawhitten
Bradston
Alternun   Dinterton
Lawannok   Myrlton
Trelaske South   Siddenhm
Pitherwin   DE
Lasant
Bradston
Lawhitten
Northhill
Stoke
Lenkenhorne
Southhill Carbolok Cullacombe
parke
Kellington Coudscoke TAVESTOK
Bickton   Cuttoole
Sct Iue   VON
Newton Sct Dornyok
Quethiok Sct Mellius SHI
Mynhenyet Pillaton Chyston
Chafrench Wotton Cargrene Bears
Lanrake See Erne Landly
Tremar... eart SALTASE
Sct Germans Wareby
Crost hole Tamerton folhe
Tregouik Sheuiok Sct budivk
Earthe
West Auton Stokedameron
Sct Merten   PLYMMOVTHE
Low   Setton key Leuſhore
Sct Michaels Insul Ef Aut... Milbrok RE
Sct Johns
Maker
Rame Edgcumbe
Catwater
Sct Michaeli tor Redford
Plymmouth hauen
Rame hede
Mewstone

PESTIS PATRIÆ
PEGRICIES

NVS

ORIENS

MAP OF ELIZABETH I'S NORFOLK PROGRESS,

William Bowles

The simple sketch map charts the intended course of Elizabeth I's royal visit (or 'progress') to East Anglia in 1578. The journey was part of the queen's regular summer pattern of touring the counties closest to London as a means of bringing royal power into her subjects' lives (or at least those of the more aristocratic persuasion) rather than allowing herself to become isolated in the palaces of the capital.

Previous monarchs had been peripatetic, especially in the earlier Middle Ages before the evolution of a complex panoply of government which required the attention of the sovereign. Elizabeth's Tudor predecessors, too, had engaged in progresses; Henry VII went on several, including a four-month circuit in 1485–6 that took him to Cambridge, Lincoln, Nottingham, York, Hereford and Bristol. Elizabeth, however, took touring to a new level, engaging on twenty-three summer progresses between 1558 and 1602 which took her from London on average about fifty days each summer.

Most progresses took place close to London. The majority led the queen no further than 65 km (40 miles) from Hampton Court, though she did on occasion stray as far afield as Portsmouth, Ipswich and Coventry. She had to take with her an enormous entourage, of cooks and seamstresses, maids and men-at-arms, together with the secretaries and ministers who were needed to ensure that the business of government was not neglected (though it inevitably ran slower on tour, leading some, such as her spy-master Sir Francis Walsingham to refuse to participate in the progresses). The whole royal caravan and its provisions lumbered along in 200 or 300 carts on roads that were often severely rutted – local parishes, who were responsible for their upkeep were only obliged to allocate four or six days annually for their repair – meaning the maximum rate of travel was only about 19 km (12 miles) per day.

Ahead of the queen's advance, court officials would long before have surveyed which houses she would stay the night in, and which she would merely stop at for dinner. Local mayors and lord-lieutenants were also required to declare if there had been any cases of the plague in their areas (in which case the progress would divert around them). Most stops were overnight and, though the hosts might be expected to provide suitable presents for Elizabeth and to permit the purchase of provisions at reduced rates by the Yeoman Purveyors of the Court, the overall cost for them was onerous, but not ruinous compared to the possible advantages gained by attracting favourable royal notice.

The Norfolk progress of 1578 was planned by William Bowles, a Yeoman of Her Majesty's Chamber, who drew the sketch map (this section showing the return leg from Thetford to Greenwich). There was serious business to conduct that summer, as the English government wanted to prevent the French from intervening in the Netherlands, where a rebellion against Philip II had been fanned by mutinous Spanish troops who sacked Antwerp. Envoys of the French commander, the Duc d'Alençon, sought audiences with Elizabeth, who was also plagued by a severe toothache all summer and cannot have been in the best of moods.

The court left Greenwich on 11 July, accompanied by 130 Yeomen of the Guard and a detachment of Gentlemen Pensioners (a mounted guard established by Henry VIII) and spent the first six nights at Havering Palace. Before she reached Norwich on 16 August, there was the normal round of visits to aristocratic houses (including that of Thomas Howard at Audley End, whose father she had had executed in 1572). There were moments less tasteful to the court; at Bury St Edmunds, Lord Burghley was so scandalized by the pro-Catholic sentiment he encountered that he accused the townsfolk of being 'affected with the brainsick heresy of the papistical Family of Love'.

Norwich, though, was the highlight. With 16,000 inhabitants it was the second city of the realm. Elizabeth was greeted with lavish entertainment; the mayor gave a speech of welcome in Latin and presented her with a silver and gilt cup worth £100, while troupes of musicians and mummers put on a show, and a huge procession led her to the cathedral, where she heard the singing of the *Te Deum*. So gratified was the Queen after her six-day stay, that at the end of it she knighted five Norfolk gentlemen.

The progress returned to Greenwich on 23 August, having stayed overnight at twenty-five places during the six weeks of the tour. Costly, lavish and time-consuming though it had all been, it was not a mere piece of self-indulgence on Elizabeth's part. As an unmarried woman with no prospect of producing an heir, she contrived to turn herself almost into a symbol of the monarchy, into Gloriana, and in this metamorphosis the progresses played a vital role.

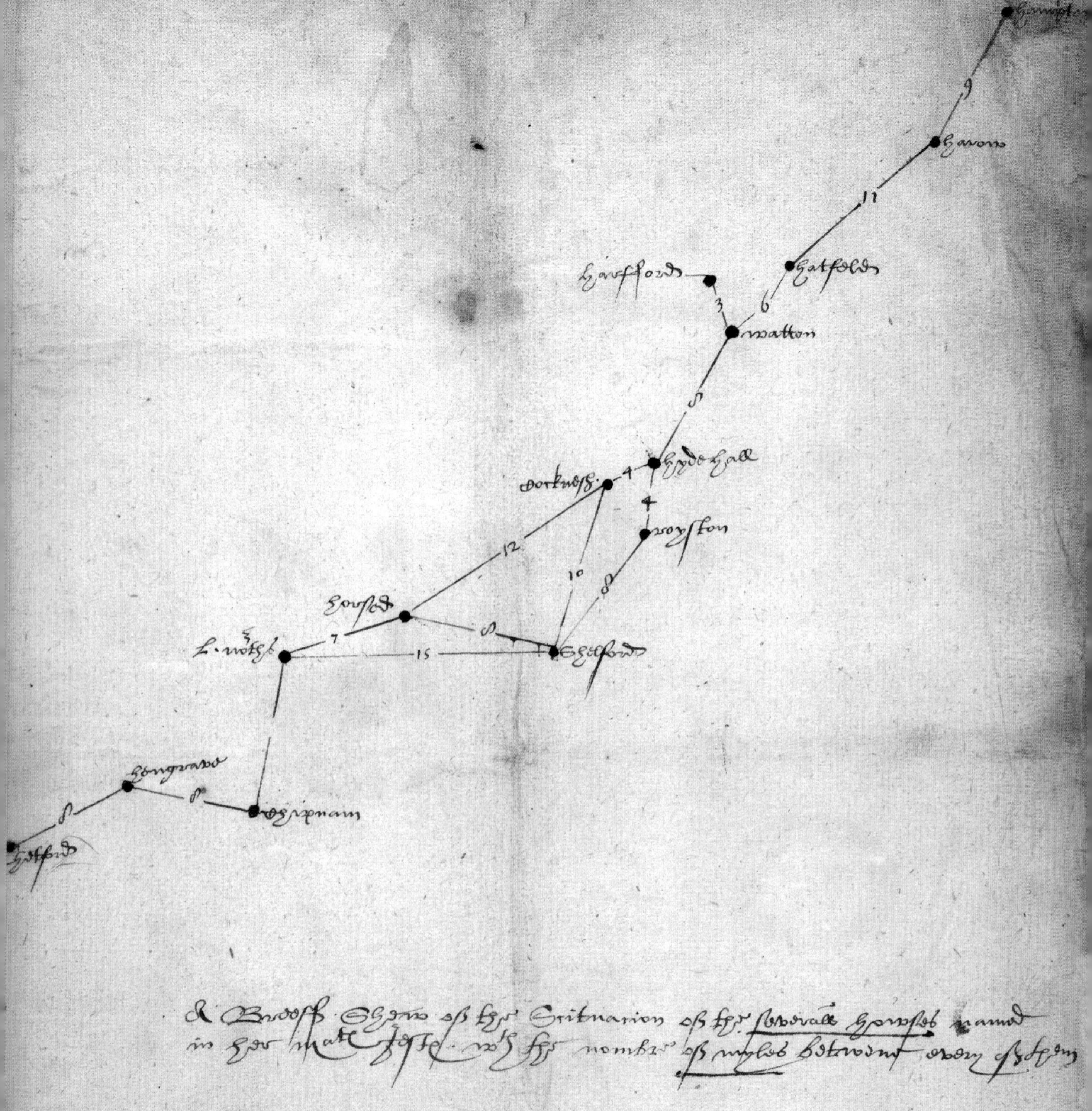

Hampton

9

heron

11

Hartford   Hatfold

3   6

watton

5

Perkness   4   Hyde hall

4

12   Royston

10   8

Horsed

L. noth   7   6

15   Stansted

Hengrave

8   8

Bishopnam

Hatfeild

A Brevff vewe of the Scituation of the severall howses named
in her mat[jes]ts [...] w[i]th the nombre of myles betwene every of them

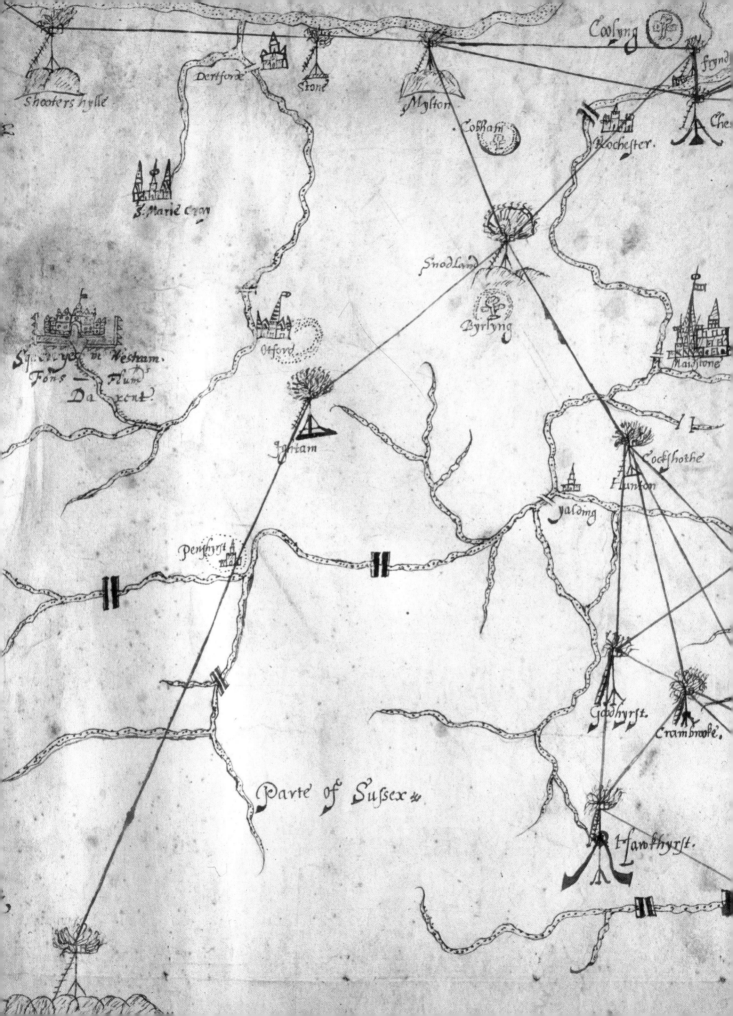

Coolyng

fynd

Shooters hylle

Dertforde

Stone

Mylton

Cobham

Rochester

Che

S. Marie Cray

Snodland

Byrlyng

Maidstone

Squeryes ne Wesham.
Fons — Flum
Darent.

Otford

Ightam

Cockshothe

Hunton

Penshyrst

Yalding

Parte of Sussex

Goudhyrst

Crambrooke.

Hawkhyrst.

BEACON MAP OF KENT,
William Lambarde

Lines radiate out in all directions from the principal town and villages of Kent in this 1585 map by William Lambarde. Yet they are not distance indicators nor aids to route-finding, but sight lines for the network of warning beacons which dotted the county, waiting to be lit the moment the watchers who manned them caught sight of an enemy.

The notion of communication beacons was an ancient one, the light of their fires travelling faster than any messenger could possibly do. The first definite mention of them in England comes from the Isle of Wight, in 1324, where thirty-one beacons were recorded as being in use to guard the island against attack. In Kent, a royal order was sent in 1326 referring to the 'said watch [who] watch have a sign of fire, or of other effective means, which can be seen from afar, so that the men of the neighbouring parts may be able to betake themselves to the fire, or other signal, in the night.'

As ever, it was fear of invasion from the continent, and principally from France, that motivated the retention and maintenance of the beacon network, a task which was all too often neglected in times of peace. In 1539 Henry VIII's minister Thomas Cromwell was forced to issue an order that the beacons were to be renewed, at a time when tensions with France were once again rising. In 1546 more detailed instructions were issued – each station was to have three beacon fires, which were all to be lit on sight of a full invasion force (but only one in the event of a lesser incursion).

The system was on occasion misused, and the rebels of the Pilgrimage of Grace, who rose up in northern England in 1536 in a bid to force Henry VIII to restore Catholicism, were mustered by means of the beacon system of Yorkshire. There were occasional false alarms, too. In July 1545, the men of Kent sent a stiff protest to London concerning the behaviour of their counterparts in Sussex who had fired their beacons warning of a full invasion that never materialized. To resolve this, it was determined that only when ten hostile French ships were definitely sighted and had begun to offload troops, should the beacons be lit.

When Lambarde's map was issued in 1585 he was forced to rebut suggestions that publishing the beacon map was actually aiding the enemy, pointing out that the merits of alerting the county militia to an invasion far outweighed the advantages to the invaders in knowing they had been seen. That enemy by now was Philip II of Spain, angered by Elizabeth I's restoration of Protestantism following the death of her sister Mary (who had been Philip's wife) and by her support for Dutch rebels against Spanish dominion in the Netherlands.

The beacons' greatest test came in 1588, when Philip II dispatched his great Armada up the English Channel to rendezvous with an invasion force waiting on the Dutch coast under the Duke of Parma. The Armada's progress was slow – it took a week to reach Calais (where it was comprehensively defeated by the English fleet under Lord Howard of Effingham and Sir Francis Drake) – and so the guardians of the beacons should have had ample time to light their fires. Yet there are no reports that they did so in Kent (and only one from Portsmouth). At England's time of greatest trial, its beacons proved to be a damp squib.

Nonetheless, the beacon network was retained and an order to upgrade it came in 1596 after the Earls of Essex and Nottingham attacked Cádiz, ransacking the Spanish port, extracting a fee of 120,000 ducats to evacuate it and taking a clutch of hostages for further ransom. Retaliation was feared, though it never came. The beacons were still being manned in 1640, but by 1672 had fallen into disuse as a scheme was proposed to revive them, perhaps as a result of one of Britain's greatest naval disasters, when a Dutch fleet crept unseen up the Medway in June 1667 and attacked the British fleet at anchor in Chatham, burning thirteen ships and capturing two including, humiliatingly, the flagship the *Royal Charles*.

The beacons' long duty as a warning of war was over, however, and they became a symbol of nostalgia for the glorious past of the Elizabethan era, finding their most recent use during the Diamond Jubilee of the second Queen Elizabeth, in 2012, when thousands were lit across the country to mark the sixtieth anniversary of her accession.

see more on next page >

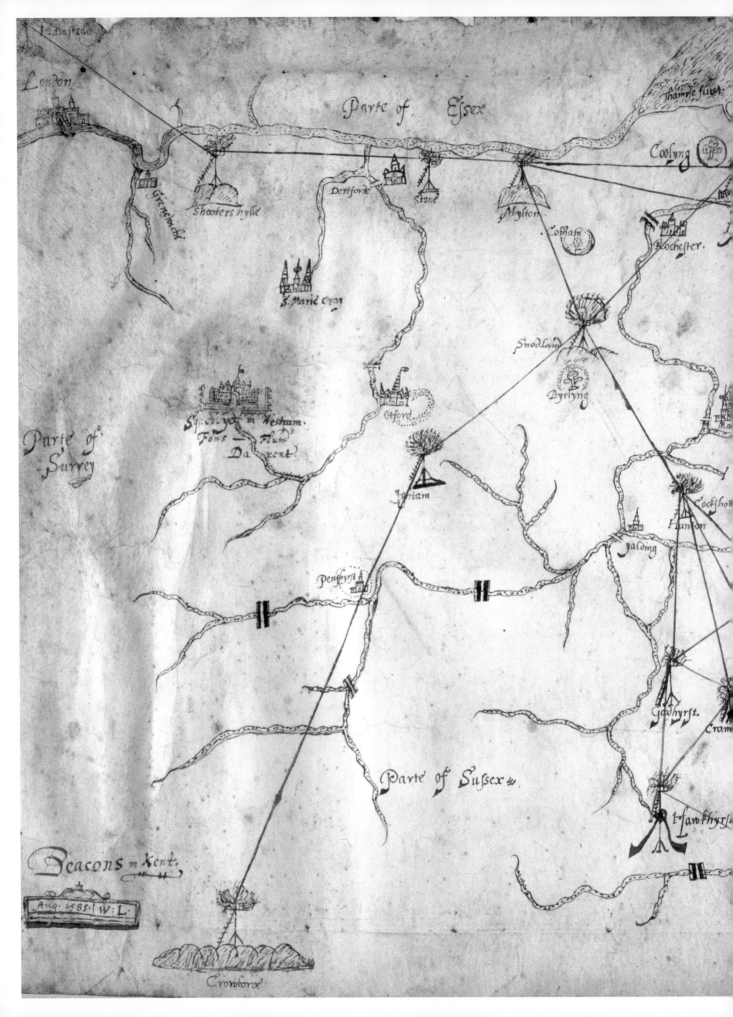

Hamstead

London

Parte of Essex

Thamfe fluvi:

Coolyng

Grenewiche

Shooters hylle

Dertforde

Stone

Mylton

CoSham

Rochester.

Sᵗ Marie Cray

Snodland

Byrlyng

Squeriel in Westham
Fons — Flum
Da Kent

Otford

Parte of
Surrey

Ightam

Cockfho
Hunton

yalding

Penthyrst

Parte of Sussex

Goudhyrst

Cram

Hawthyrst

Beacons in Kent.

Aug. 1585. W: L:

Crowforde

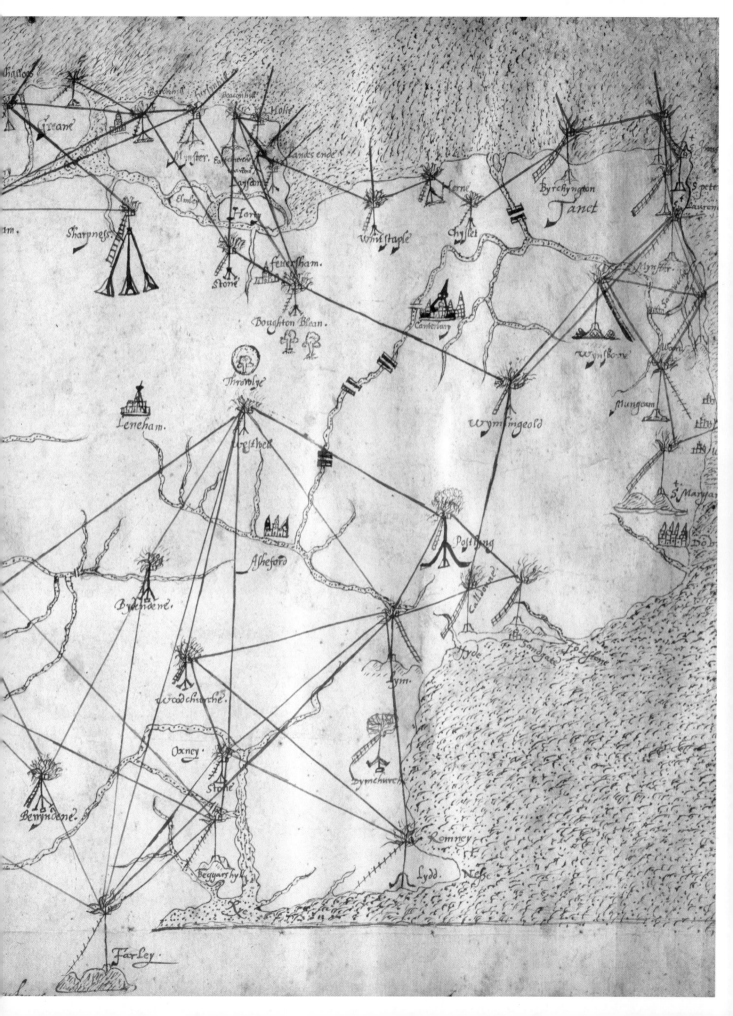

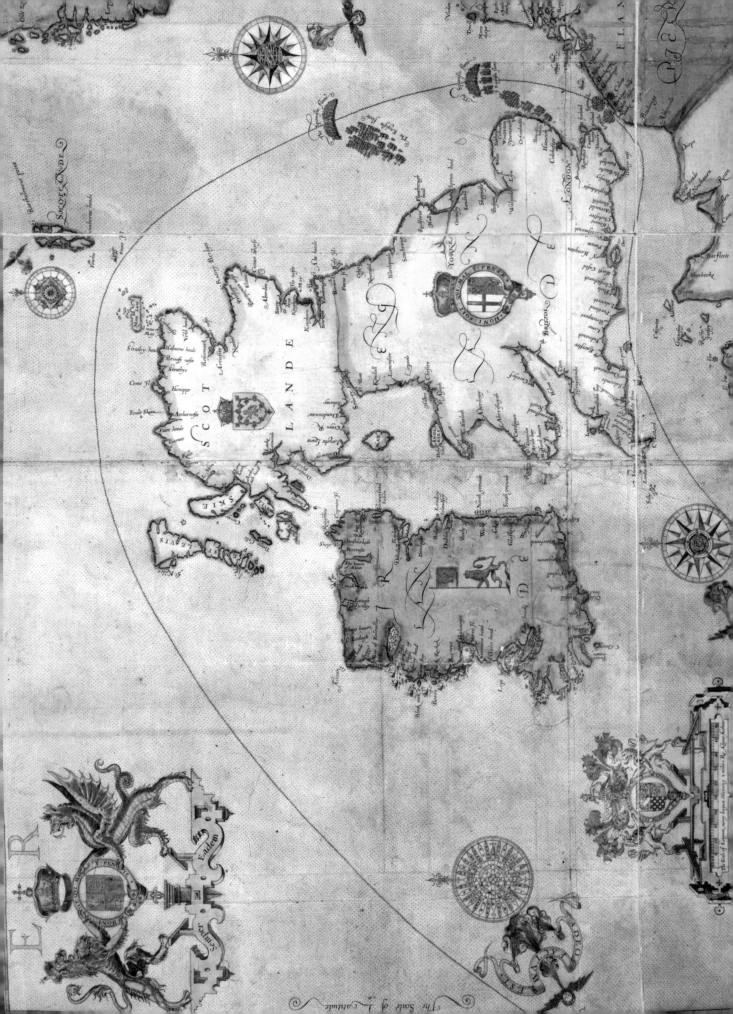

MAP OF THE SPANISH ARMADA, Robert Adams

A thin line looping around the east coast of England and the west coasts of Scotland and Ireland in this 1588 map by Robert Adams marks a moment of national crisis. It traces the route of the *Grande y Felicisima Armada* ('Great and Most Glorious Armada') despatched by Philip II of Spain to punish Elizabeth I for her support for Protestant rebels in the Netherlands and her sponsorship of privateering raids by Sir Francis Drake against Spanish ports in the New World and Spain itself.

Adams, who was Surveyor of the Queen's Buildings, produced his map in an atmosphere of euphoria and relief that followed the defeat of the Spanish Armada. It illustrated a book by Petruccio Ubaldini, a Florentine calligrapher and diplomat, who saw the commercial possibilities of a piece of populist propaganda and produced an account of the Armada campaign, based on interviews with Drake, the equally self-promoting hero of the hour.

Adams tactfully pays visual homage to two more powerful patrons: Lord Howard of Effingham, the Lord High Admiral, who commanded the English fleet (and whose coat of arms sits at the bottom left of the map) and Elizabeth I, whose personal standard (together with her motto *Semper Eadem*, or 'Ever the Same', a reference to her faith and constancy) is positioned at the top left (the map shown has been rotated through 90°).

It was Howard's responsibility to impede the progress of the Armada through the English Channel, a task made more difficult by its sheer size (with 129 vessels of which 35 were large warships) and to prevent it rendezvousing with a 26,000-strong invasion force mustering in the Spanish Netherlands under the Duke of Parma. In Howard's favour were the comparative manoeuvrability of the ships under his command as compared to the lumbering Spanish galleons, and the character of his opponent, the Duke of Medina Sidonia, a bureaucrat with no naval experience who had only received the commission when his predecessor most unfortunately died of typhus.

Partly scattered by storms in the Bay of Biscay after their departure from Corunna on 20 July 1588, the Spanish fought running skirmishes with the English along the length of the channel, including with Drake who had, the legend grew up, rather grumpily insisted on finishing the game of bowls he was playing on Plymouth Hoe when news of the Armada's approach was brought to him. By 27/28 August the Spanish fleet was mustered at Calais ready to escort Parma's invasion barges over to the Kent coast. Then, Howard struck a devastating and entirely unexpected blow against them. He sent eight ships, abandoned hulks packed with an incendiary mix of rags and tar, careering towards the Spanish fleet.

The fireships were set alight and panic spread among Medina Sidonia's captains. They surged out of the safety of Calais harbour straight into Howard's waiting formation. Five sank or foundered almost immediately; the rest found their way blocked by the English fleet and impeded by winds which left their only avenue of escape northwards into the North Sea. So the Spanish fled, and continued to flee. Medina Sidonia paused only to throw surplus horses overboard and to hang the captain of the *Santa Barbara* for cowardice. Howard followed them all the way to the Firth of Forth until, on 12 August, it was clear that they no longer posed any threat and he broke off the pursuit. For the Spanish, though, the agony was not yet over. Medina Sidonia ignored the advice of senior captains that, given the Armada had no accurate charts of the coast of Scotland and Ireland (having expected never to need them), he should give these a wide berth.

Storms and navigational errors led to dozens of Spanish ships sinking or running aground. Many of the survivors were summarily executed by the English authorities in Ireland. Around 300 crew and soldiers aboard the *Valencera* were mown down by arquebus fire in a Donegal field after their ship ran aground; only the forty-seven officers were saved for ransom. By the time the survivors of the fleet limped into Coruna and other north Spanish ports in September and October, it was missing half the vessels which had set out with such high hopes.

In England there was jubilation as news of the Armada's defeat began to trickle through. Celebratory bonfires were lit throughout the south, captured Spanish battle standards were paraded at St Paul's and Elizabeth gave a resounding speech at Tilbury declaring 'I know I have the body but of a weak and feeble woman; but I have the heart and stomach of a king, and of a king of England too.' The myth of English naval superiority was established, too, which would sustain the nation through Trafalgar and into the First World War. It was the idea – that the sea was England's shield – that Adams' map played a key role in sustaining.

LEMMINGTON

VERTON

EDMONDSCOT

GYES
CLIFFE

WOODLOV

WARWICKE

WEDGNOKE
PARKE

BVDBROKE

TAPESTRY MAP,
commissioned by Ralph Sheldon

Richly woven in silk and wool, Richard Sheldon's 1590s tapestry map of Warwickshire is one of the most arresting cartographical representations of England from that time. Its subdued hues and wealth of depictions of the county's forests, hills, churches and towns create an impression of a nation rich and at ease with itself. As a piece of visual propaganda it was most effective, in an age when the harnessing of such techniques to create national myths was ever more important.

The Elizabethan age was a propaganda age, as plucky little Protestant England fought alone – or sometimes not quite so alone, as it often allied with the French – against the dark forces of Spain and Catholicism. There was much in this myth that was true, but the creation of a common national narrative did wonders for morale, especially in a nation that within the past century had been buffeted by the religious winds of the Reformation and torn asunder by the dynastic carnage of the Wars of the Roses.

Elizabeth was fortunate that her victories against Spain, most notably against the Armada in 1588, gave her adoring subjects something to coalesce around. With the victory over Spain, Elizabeth freely displayed the world map which Sir Francis Drake had presented to her following his circumnavigation of the globe in 1580. It may have attracted the ire of the Spanish – into whose lands Drake (given by them the half-admiring nickname 'El Draco', the dragon) had trespassed and who continued to launch provocative raids there – but it certainly got attention, and Henry IV of France ordered his own personal copy.

The Armada victory was chosen as the subject of a set of tapestries commissioned by Lord Howard of Effingham, Elizabeth's Lord High Admiral, and which hung on the wall of the House of Lords until their destruction in the fire which destroyed the Houses of Parliament in 1834. Tapestries were all the rage both as a means of presenting an (often pointed) visual message, but also as a display of wealth. The finest weavers came from Flanders, and their skill and finesse meant that the more than 100 who migrated to England in the late sixteenth century to escape wars at home found ready employment.

One man who established a weaving business, at Barcheston in Warwickshire, was Ralph Sheldon's uncle William. From 1570 his works produced a stream of products to satisfy the thirst for Flemish-inspired tapestries. They may have been responsible for the four huge tapestry maps ordered by Ralph around 1590, depicting the counties of Gloucestershire, Worcestershire, Oxfordshire and Warwickshire and intended for his new house at Weston in south Warwickshire. They were heavily based on Christopher Saxton's new county maps, though at 6 m by 4 m (20 ft by 13 ft) there was ample room to include more details of the landscape, including drawings of the principal towns (such as Warwick, where Warwick Castle and the tower of the Collegiate Church of St Mary can be seen). Not everything was accurate though; in deference to his family interest in the tapestries, Sheldon had his Weston house shown as larger than the nearby village of Long Compton.

At the top right of each map, paraphrases of passages were included from William Camden's *Britannia*, the first antiquarian study of the history and customs of England's regions, which was newly published, having been completed in 1586. And at the top left, Sheldon had the coat of arms of Elizabeth I embroidered into the tapestry. It was a diplomatic acknowledgement of the ageing queen, who had now had become enshrined in the national consciousness as the Virgin Queen and Gloriana (a character in Edmund Spenser's moral allegory *The Faerie Queene*, which was intended as a scarcely veiled reference to Elizabeth). Sheldon's tapestry map is very much in the spirit of the age, an age in which even tapestries could be harnessed to the greater glorification of the sovereign and of the nation.

see more on next page >

Scala paſſuum.
80   160   240

More feyldes

Spittle feyldes

Moregate

Biſhopeſgate

S. Botolph

Aldgate

Crowne poſterne

More feyldes

Three cranes

The Billarde

Shrewſburye howſe

Olde ſwann

Lion kaye

Bellyns gate

Galley kaye

Cuſtom howſe

The to

S.

fluuius

S. Marye Oueryes

S. Towleyes

MAP OF LONDON, John Norden

John Norden's 1593 map of London portrays a city in transition, during the period when it was beginning to burst out of its medieval bounds and its ancient Roman walls, and at a time when a newly self-confident England was reaping the rewards of thirty-five years of Elizabeth I's careful and painstaking re-establishment of its political stability.

The map's label reads that it is 'A guide for Cuntrey men in the famous Cittey of London', and its maker, John Norden, had provincial origins, coming from Somerset before he made a name for himself as a surveyor in royal service. In 1593, he embarked on a large-scale project to produce a series of county maps and histories. The first part of this *Speculum Britanniae* ('Mirror of Britain') covered Middlesex (including London) and was dedicated to the Queen in the hope of receiving further royal patronage. However, only three more counties (Essex, Northamptonshire and Cornwall) were published before Norden's death, as the hoped-for subventions from the court soon dried up.

Norden's London (in this reprinting from 1653) shows its ancestry, with the twisting network of medieval streets crowded inside the old Roman walls (punctured by its ancient gates, such as 'Moregate', 'Aldersgate' and 'Ludgate'). All around are fields, and areas now in the capital's urban core, such as Islington, are still mere villages. The inner area was becoming ever more overcrowded as the population of London rose from around 50,000 in 1520 to four times that number by the end of Elizabeth I's reign in 1603.

Along the banks of the Thames, Stow labels prominent buildings. In the somewhat seedy entertainment quarter of 'Southwarke' he places 'The Beare howse', a venue for the perennial Tudor favourite sport of bear-baiting (and a replacement for a previous building which had collapsed in 1583 when its rotting timbers gave way, killing eight spectators). Nearby is the 'The play howse' – the Rose, which was Southwark's first purpose built theatre to cater for newer and more refined tastes.

On the Thames's north bank is 'Baynardes castle', a veritable microcosm of English history. At a fort here the Danish King Canute is said to have spent the winter of 1017, on a site which became a Norman fort after the conquest of 1066. It later came into the possession of Robert FitzWalter, the standard-bearer of baronial opposition to King John in the years before the Magna Carta. After a stint as a Dominican priory, it became an aristocratic residence and saw the proclamation of Richard III as king in 1483 and that of Queen Mary (and the deposition of her unfortunate cousin, Lady Jane Grey) in 1553.

By the time Norden drew his portrait of the city, the country had recovered from both from these trials and the euphoria of the victory over the Spanish Armada of 1588. But the Queen's reign was winding down, as she lost her youthful steel and fell under the shadow of a series of favourites. The last of these was the dashing Robert Devereux, Earl of Essex, who dazzled his ageing monarch and inveigled her into allowing him to take part in a revenge raid against Cádiz in 1596, as part of a diversionary move to draw Spanish troops away from England's hard-pressed allies, the Dutch and French. Cádiz was captured, but promptly handed back, much to Essex's chagrin. He used the political capital he had gained to bully Elizabeth into allowing him to lead a campaign against Irish rebels in 1599, but he soon tired of the fighting and instead of subduing the Earl of Tyrone, the rebel leader, he negotiated a truce with him. An infuriated Elizabeth had him arrested and confined to his house on the Strand.

Bereft of royal favour, and in deepening financial difficulties, Essex hatched a plot to seize back his position at Elizabeth's side, but, vain as ever, he overestimated his own importance and the mob of 300 supporters he led down Cheapside on 8 February 1601 chanting 'The Crown of England is sold to Spain' were met with mute indifference. Essex was soon apprehended, summarily tried and condemned. Elizabeth stayed his execution several times, but after a long evening at the theatre changed her mind and gave the final word. On Ash Wednesday, 25 February 1601, Essex was hung at the Tower.

Two years later Elizabeth, too, was dead without having given any idea of who might succeed her until she was on her deathbed, when she indicated that it should be James VI of Scotland, son of her cousin Mary, Queen of Scots, who then began the long journey down to London. An era was ending, both for London, and for England, which was soon to be joined in personal union with its Scottish neighbour as the island of Britain became united for the very first time.

see more on next page >

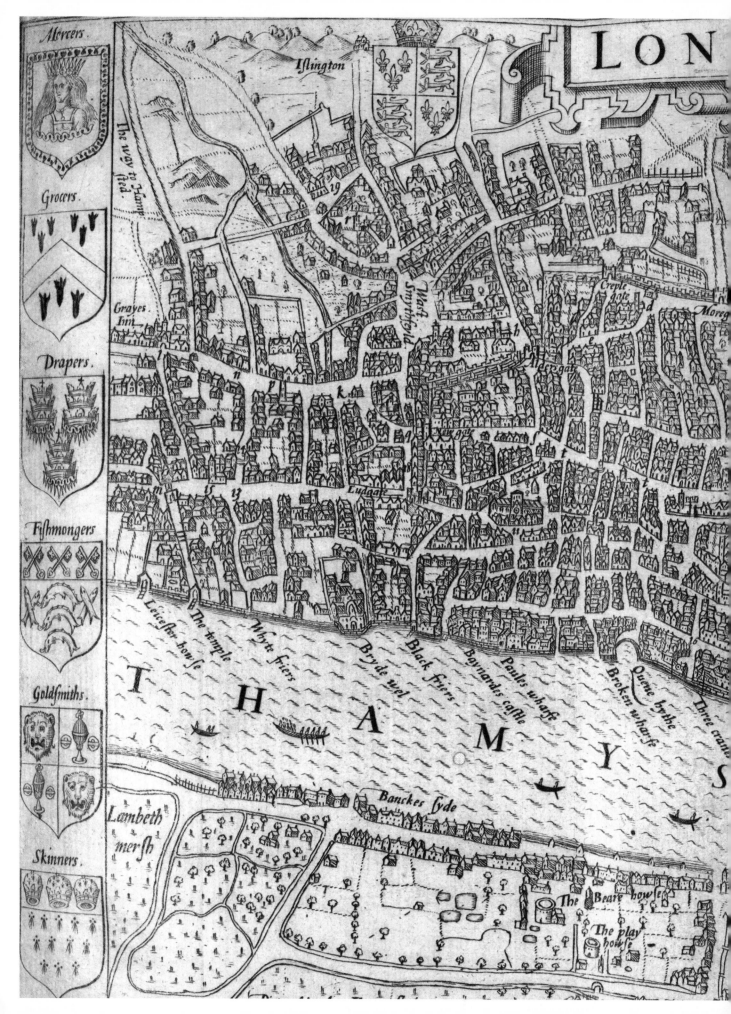

LON

Mercers.

Grocers.

Drapers.

Fishmongers

Goldsmiths.

Skinners.

Islington

The way to Hampstead.

Grayes Inn

Creple gate

Moreg

Weast Smythfeld

Aldersgat

New gate

Ludgat

Leicester howse

The Temple

Whyte Fryers

Bryde wel

Black Fryers

Baynardes castle

Paules wharfe

Broken wharfe

Quene hythe

Three cranes

T H A M Y S

Banckes syde

Lambeth mersh

The Beare howse

The play howse

ON

Merchantaylo[rs]

Haberdashers

Salters.

Ironmongers.

Vintners.

Clothworkers

Scala passuum s. pedum

80   160   240   320   400   480

Spittle feyldes.

S. Botolph

Aldgate

Bishopes gate

Spittle feyldes.

East Smithfeild

towre
poster
ne

Olde swan
Shrewsburye howse
Lion kaye
S. Bellyns gate
Galley kaye
Custom howse
The towre
S. Katherynes

fluuius

Marye
oueryes
20

S. Towleyes

South warc

21   22

SPECULUM BRITANNIAE,

John Norden

The division of the county of Surrey into hundreds, and the inclusion of a scale and a key, mark gentle advances in the art of English cartography in this 1594 map of Surrey by John Norden. Trying to improve the county maps of his predecessor, Christopher Saxton, Norden worked right at the end of the Elizabethan age, a time both of confidence – now that the threat from Spain was receding – and concern for the future, as the Queen, ageing and heirless, had only a little time left to live.

Born in Somerset, Norden came from a gentry family, but turned his hand to estate surveying and cartography. His grand project was the *Speculum Britanniae* ('mirror of Britain') conceived as a pocket-sized county-by-county guide to England, to be accompanied by both maps and a text recounting each area's history. Norden began work on it in 1590, but only ever completed the full volumes for Middlesex and Hertfordshire, before the death of his patron Lord Burghley in 1598 caused the flow of funds to dry up. For most of the counties, including Cornwall, Hampshire, Sussex and Surrey (the last finished in 1594), Norden completed only the maps. For the rest of his career he relied upon his surveying for an income, culminating in his appointment in 1612 as surveyor of the royal castles in ten counties in southern England and – when that was not enough – on the money derived from his writing of popular devotional tracts, including *A Pensive Mans Practise*, published in 1584, which went through forty editions in his lifetime.

Norden's Surrey map included the whole historic country, which before the local government reorganization of 1889 extended to the banks of the Thames at Southwark. It was a land steeped in history, settled before the Roman conquest by the Atrebates tribe and tranquilly passing through four centuries of Roman occupation before it was overrun in the first wave of Anglo-Saxon conquest in the late fifth century. Among its principal settlements was Guildford (Guldford on the map), the future county town of Surrey, which first appears in written records in 880 in a charter of Alfred the Great (in which it is named as Gyldeforda).

In 1036, Guildford achieved a more sinister reputation as the site of a notorious massacre. King Harold Harefoot, the Danish ruler of England (and the son of Canute) had acceded to requests by his stepmother Emma that she be allowed to see Alfred and Edward (her sons by Canute's predecessor, Aethelred the Unready). Alfred landed in Sussex and made his way overland towards London. When he reached Guildford he was met by Earl Godwine, the king's trusted adviser (and father of another King Harold, who would die at Hastings in 1066). When night fell, Alfred's Norman mercenary bodyguards were dragged from their beds and brutally slaughtered. The prince was only spared long enough to be taken to Ely Abbey, where he was tortured, blinded and shortly after died. Had he not fallen into the trap, it would have been he, and not the pious Edward the Confessor who became king in 1042.

Following the Norman conquest, Guildford remained a focus of royal attention, as William the Conqueror had a castle built there just after 1066. King John visited no fewer than nineteen times, the last in 1216 after the barons had forced him to sign the Magna Carta and, having tried to renege on the deal, he was faced with an invasion by the dauphin of France to force him to accede to its terms. Shortly thereafter, the town acquired an interest in the Crusades when the estate of Merrow (Meroe on the map), to the east of Guildford, was granted to the Knights Templar. A crusading order, and amongst the most effective defenders of the holy places of Palestine, the Templars had a network of knights, supporters and landholdings to help fund their operations and raise recruits when needed. Unfortunately, their wealth and secretiveness drew accusations of heresy from jealous opponents and in 1307 the order was suppressed in France. By January 1308 English Templars had been arrested, but most were treated leniently and only one was executed. They were lucky to escape with the confiscation of their lands, much of which was simply transferred to the other main crusading order, the Knights Hospitallers, as their sister branch of the Templars in France was brutally suppressed, with many of its leaders being burnt at the stake in 1314.

Had Norden completed his *Speculum Britainiae*, he doubtless would have included these tales and more, including the career of William Shakespeare, England's leading playwright, whose Globe theatre opened in Southwark (then in the far north of Surrey) in 1599. Instead, shorn of its planned text, the map is a very indistinct mirror of the Britain of the time.

LONDON

Blackwell
Southwarke · Rcdith
Chelsey · Lambeth · Hatcham
Fulham · Newingtoñ · Peckham
Battershey · Walworth · Detforde
S. Lambeth · Camervell · Grenewic
Brantforde · Barns · Berneluces · Putney · Clapham · Ketsbroke
Heston · Syon · Kew · Mortlake · Waudesworth · Stokewell · Peckham Rey · Ley
Cranforde · Shyn · E.Shene · V · Knightes hill · Dulvich · Lewesham
Bedfont · Humslo · Thistleworth · Richmond · Rowhampton · Stretham
Ashforde · Hamworth · Combe parke · Towting gra: · Pensgreene · O
Staines · Astlcham · Twickenhā · Petersham · Towtingbeke
Egham · Littleton · Kenton · Tuddington · Wimbledon · Mycham · Croydon · Addscombe
Backham · Lalam · Charleton · kingston · Martyn · Wallington · Halug · Addington
Backhamlane · Thorpe · Sunbury · Hamton · Combe Neuell · Moredon · Bedington · VI
Stroude · Chersey · Shepperton · Court · Maldon · Sutton · Carshalton · F.
S. Ann hill · Addstone · Abscourt · IIII · Long Ditton · Cheam · Woocote · Saunderste · Chelsham
Newlodge · Walton · Mowlsey · E.Mowlsey · Thams Ditton · Woodmansturne · Warlingham · Woldinge
tions ponde · I · Otlandes · Esher · Vleagate · Talworth court · Nonsuch · Chipsted · Couldesdon · Morden
Woodham · Waybridge · S.Georghill · Esher · Cherneworth · Chesington · Ewell · Bansted · Chaldon · Katerham
Horsyll · Lane · Wisley · Abroke · Horton · Burghhouse · Mestham · Willye · Hackstall
Chobham · Rysley · Byflet · Cobham str: · Ashsted · Ebesham court · Tadworth · Gatton · Godstone
Purforde · Cobham · III · Ebesham · Walton · Wiggy · Nutsfyld · Blechingligh
Wokinge · Newarke · Stoke Dabern · IX · Hedley · Buckland · Lyngsfeld str · Poundhill · South pke
Bridley · Ripley · Okham · Slyfeyld · Norbury · Mychelham · Colley · Reygate · Reygate church · Bustow pke
Mayforde · Sende · Great bookham · Feeham · VIII · Woodhatch · Douers
Sutton · II · W.Horsley · Preston · Little bookhā · Brockham · Flanchford · Kynnersley · Hurne
Burphants · E.Horsley · Essingham · Polsdon · Beachworth · Lee · Burstowlodge · Smalsfylde
stowghton · Stoke · W.Clandon · Robarns · Darkking · Beachworth · Sydlune myll · Laystwte · Bish courte
ord maner · Guldsford · E.Clandon · Wolton · Mylten · Eywood · Shelwood · Horley · Hodge court
Meroe · W.Clandon · Wescot · X · Buckland · Layw · Shipley br · Imberhor
erm hill · Shere · Holmewoodboro · Brockham · Crabbet
Littleton · Tyting · S.Martins · Abinger · Chappell · Tempell · Newdigate · Ifeyldcourt pke · Little
thesley · Shalforde · Chilworth · Gumshall · Haw · Charlewood · Worth · Woreth
Vsted · Tangley · Ognersh · Weston · Holmbury · Lethe · Okely · Ifeylde · Ewhurst · Crawby
Godalning · Bramley · Winterfolde kill · Ruspey · Bonnykes · Beawbushe · Tilgate · Forest
Catteshill · Mounsted · Ewhurst · Shelky
Burgate · Halcombe · Scotsland · Cranley · Oke woode · Warncham · Reffey · Balcombe
enton · Fann · XI · Knoll · Nore · Vachery · St
eldon · Loxley · Baynards · PARTE OF · Horsham · Leonardes · Slawgham · Sydney
Dunsfolde · Cranley · Detham · Handcrosse · Tyes
Burningfolde · Glashowse · Aurfolde · Simfolde · Stamerham · Chesworth · Slohowse
Shillingleigh · Sydney · Rydgwycke · Itchingfold · Nuthurste · forest · Holmsted · Brandeñ
Loxwood

Pet pum Hov Syidlovill Balkach

S ogil Balydlouum

Dronla Strykmartin My

ca: S.

Affry Kirktou of strichmartin Balmuchil houses

Pietstoun Baldeue Kotte

Kllauerhous

Balera g S graha Kirktou Muidieu

Clepingtou

Widdle poundy Westmoore Ret

E: pouirdy

Pateloy Drybrigh Kirktou

Duntay W: clepingtou halof fullains

idy DVN-TAY Mains

K: of Mooreknoues

Capelbeun Kragt

New biggin

Blackness Milto

Bolga Groters

Butt bed steun

Warig

MAP OF SCOTLAND,
Timothy Pont

With its dense clustering of towns and villages, and its meticulous portrayal of ports, market-places, city-walls and churches, this late-sixteenth century map by Timothy Pont of Lower Angus and Perthshire east of the Tay forms part of a set that make up the earliest accurate map of Scotland. It portrays a growing and prosperous nation, just on the edge of an enormous change that was to see it lose its separate political identity.

Pont was a mapmaker from Fife, whose unassuming career as a church minister in Caithness left little trace in the records. Yet the 130 maps he produced in the 1590s – the only securely dated one is for Clydesdale in 1596 – are the finest contemporary visual record of Scotland and, over forty years after his death, they formed the basis of the Scottish section of the great Dutch cartographer Johannes Blaeu's 1654 world atlas, the *Theatrum Orbis Terrarum*.

Scotland's population was growing during Pont's lifetime. While in 1500 the population was around 700,000, it had burgeoned to a million in 1700, despite outbreaks of bubonic plague in 1568–9, 1574 and 1584–9. New towns were growing to challenge the centuries-old primacy of Edinburgh. Dundee on the Tay, in the central section of this map, took advantage of its port (lovingly detailed in Pont's engraving) to become the country's second city by the sixteenth century; in 1495 its assessment for taxation was £250, second only to Edinburgh (and a nose ahead of Aberdeen, which paid a levy of £200).

Scotland had experienced a turbulent time in the decades before Pont's map. James VI was an infant of just 13 months old when he succeeded his mother, Mary, Queen of Scots, after her forced abdication in July 1567. For the next 16 years he was dominated by a series of regents, the first two of which, the Earl of Moray (Mary's half-brother) and the Earl of Lennox were assassinated as part of a series of vicious factional struggles between the pro-Catholic party of his mother's former supporters, and an opposing Protestant and pro-English cabal. In 1582 the latter kidnapped the king in the Raid of Ruthven, and held him hostage until he escaped in 1583, and publicly declared himself to be Scotland's sovereign lord.

The young James faced the difficult task of balancing good relations with England without appearing to be too weak (his mother was, after all, still a prisoner of Elizabeth I). However, even when Elizabeth executed Mary in 1587 (after she had been caught in intercepted letters plotting her cousin's overthrown in the Babington Plot), he did little but make a token protest. At home James was kept more than occupied in a struggle with the Scottish Kirk, which under the guidance of the radical Calvinist Andrew Melville issued the *Second Book of Discipline* in 1578, eschewing bishops and denying the right of the king to intervene in church affairs.

James retaliated by having the Scottish parliament pass a series of laws in 1584 which confirmed the appointment of bishops and forbade groups of ministers to assemble in convocations except by royal permission (a clear attack on extreme Presbyterians such as Melville). In other ways, too, he proved himself an ardent exponent of royal power. Among the seven books he found time to write was the *Trew Law of Free Monarchies* (1598) which asserted that the right of kings to rule was a divine gift, which none of his subjects had a right to interfere with. It was, unfortunately, a lesson, which his feckless son, the future Charles I of England, took too much to heart in his struggles with the English parliament in the 1630s. A darker side of James's personality was revealed by his interest in witchcraft, provoked by a storm which almost sank his ship in 1589 on the way back from Scandinavia after his marriage to Anne of Denmark. Dark powers were assumed to have been involved and over 100 people were accused of complicity, tried, and the bulk executed. The king's subsequent authorship of the *Daemonologie*, a tract on witchcraft, led to a further upsurge in witch-trials in 1597.

James spent the 1590s biding his time and cultivating relations with Sir Robert Cecil, Elizabeth I's chief minister, knowing that, in the absence of the queen producing an heir (which by now was unlikely, as Elizabeth was by then over sixty years old), he might well – as the descendent of Margaret Tudor, the daughter of Henry VII – be next in line for the English throne. His patience proved well-founded, for when Elizabeth lay dying on 24 March 1603, she gave the long-awaited sign that James should succeed her. A messenger rode post-haste from London to Edinburgh, covering the ground in just two days and then on 5 April, James bade farewell to the Scottish court and took the road south to assume the English crown. The Scotland which Pont had drawn in such extraordinary detail would never be the same again. That James only ever visited his homeland once more in his lifetime showed that the centre of gravity of the new kingdom of England and Scotland most certainly lay in London.

see more on next page >

DIEU ET MON DROIT

PART OF SCOTLAND

PART OF ENGLAND

THE KINGDOME OF IRLAND

Devided into severall Provinces, and the
againe, devided into Counties.
Newly described.

The Gentleman of Ireland | The Gentlewoman of Ireland
The Cevill Irish man | The Cevill Irish woman
The wilde Irish man | The wilde Irish man

MAP OF IRELAND, John Speed

The neat division of Ireland into counties on John Speed's 1610 map belies the island's troubled past. The imposition of such administrative divisions had come only after more than four centuries of interventions by English kings to subdue their neighbours across the Irish Sea.

Speed was one of England's leading cartographers, a Cheshire-born tailor whose membership of the Society of Antiquaries gave him access to such eminent historians as William Camden (the author of *Britannia*) and a huge library of books and manuscripts. His appointment to a sinecure position at the Royal Customs allowed him the leisure and money to pursue his mapping interests. His *Theatre of the Empire of Great Britaine* was the first real atlas to cover the whole of the British Isles, a huge enterprise which took almost twelve years to complete.

Speed's drawing of a 'gentleman of Ireland', 'civill Irish' and 'wilde Irish' at the margins of the map betray an easy sense of superiority which translated on the ground to a centuries-long state of antagonism between the Irish and their English overlords (mediated occasionally by great Anglo-Irish families such as the Butlers and Talbots which had to some extent accommodated themselves to Gaelic political structures and customs).

Ireland's entanglement with the English had begun in 1170 when Richard, Earl of Pembroke (nicknamed 'Strongbow') sailed over at the invitation of Dermot MacMurrough, King of Leinster, and ultimately became so successful (ascending the throne of Leinster himself) that a jealous Henry II intervened and carved out an English territory in the east and north of Ireland. The fractious and divided nature of the native Irish kingdoms meant that, although the English settlement around Dublin (known as the Pale) had been reduced to a few hundred square miles by early Tudor times, the Irish could never wholly dislodge the invaders.

Serious revolts such as that of the Earl of Kildare, in 1534 threatened to do so, with the fury of the Roman Catholic rebels now increased by accusing Henry VIII of being a heretic, after his break from Rome. But Kildare was defeated and hung, drawn and quartered together with five of his uncles, while Henry, showing little inclination for general clemency, changed his title from Lord of Ireland to King of Ireland and forced all the Irish lords to surrender their land, whereon it was all granted back to them, so acknowledging him as their direct overlord.

Resistance to the English broke out periodically in Elizabeth I's reign, including an uprising by the Earl of Desmond in 1579–83. Far more serious was that led by Hugh O'Neill, Earl of Tyrone, whose victory over an English army at Yellow Ford in 1598 threatened briefly to crush Tudor power in Ireland for good. Elizabeth's first attempt at recovering her position failed miserably when her favourite, the Earl of Essex, negotiated a truce rather than fight O'Neill. The arrival of a Spanish army which occupied Kinsale in September 1601 then raised the alarming prospect of a foreign Catholic power poised to strike at the west of England.

In the end, Elizabeth's next Lord Deputy of Ireland, Baron Mountjoy, proved far more able. He conducted a scorched-earth campaign to deny the Irish rebels supplies and routed them near Kinsale. The Spanish fled back home and, although O'Neill and his chief Irish ally Rory O'Donnell, Earl of Tyrconnell, fought on for a further two years, they were eventually captured and the rebellion collapsed. Their escape four years later aboard a French ship never brought the succour to the cause of Irish independence that its planners had hoped, but it so infuriated James I that he sanctioned the transfer of large numbers of English and Scottish colonists to Ireland, so beginning the 'Ulster Plantations' that were intended to shift the demography against the Roman Catholic Irish and in favour of Anglo-Irish and Ulster Protestantism. It is in a way part of this project of absorption of Ireland into the English mainstream that leads Speed to portray the island as almost a mirror of England, all that is, save in his half-romantic, half-disdainful inclusion of the portrait of the 'wilde' Irish man and woman.

MAP OF SCOTLAND, John Speed

John Speed's colourful 1610 map of Scotland shows the country at a pivotal moment in its history; just seven years after James VI had taken the road south to assume the throne of England, as James I, joining both kingdoms in a union that has lasted through religious strife, civil war and political dissent since then.

Cheshire-born tailor and cartographer Speed included the map in his *Theatre of the Empire of Great Britaine*, which also encompassed country maps of Ireland, England and Wales, and many larger-scale county maps. Arrayed around its edges are the new royal family of Britain: James himself; his wife Anne of Denmark; his popular elder son Henry (who died tragically of typhoid in 1612) and his younger son Charles (who would be executed in 1649, the victim of his own intransigence in respect of the English parliament).

James's (and Scotland's route) to the union of the kingdoms was a winding one, closely intertwined with the progress of the Reformation. Religious change had come relatively late to Scotland, but when it did, it appeared in a form more radical than England. Fired by preachers such as John Knox, Scottish Protestants rejected the episcopacy (in which bishops governed the church), and by the First Covenant, drawn up in 1557, pledged to 'establish the most blessed word of God' in Scotland. This put them at radical odds with their Queen, Mary, who was a devout Catholic, and who, aged fifteen, married the Dauphin of France in 1558. By the time she returned from Paris in 1561, after her husband had died, the Reformation Parliament had swept away the vestiges of Catholicism, abolishing the ecclesiastical hierarchy and forbidding the hearing of the Latin Mass.

Capricious and badly advised, Mary did little to help her own cause. She married her cousin Lord Darnley in 1565, but shortly after the Queen began to suspect him of having an affair, he was found dead, probably strangled, but with his body undamaged, next to his Edinburgh house which had been mysteriously blown up. When Mary then married the Earl of Bothwell, the Scottish Protestant lords had had enough, and she was deposed in favour of her thirteen-month old son, James (mirroring her own accession in 1542 as a week-old baby). Once he had escaped the bonds of a minority dominated by tussles between Catholic and Protestant factions, James sought to enforce a milder religious regime, seeking to force through the restoration of the bishops in 1584.

This, and his departure for England in 1603, alienated him from many of his leading subjects, a situation made worse by his refusal to allow the General Assembly of the Church of Scotland to meet. The gulf grew wider under Charles I, a devout Anglican who in 1625 ordered the restoration of church lands and who in 1633 insisted his Scottish coronation (in St Giles's Cathedral in Edinburgh) be held with full Anglican rites. The imposition of a version of the *Book of Common Prayer* in 1637 led to riots and the adoption the next year of the Covenant, a protest against the perceived restoration of Catholicism, which spread widely throughout Scotland.

As a result, when Civil War broke out in England in 1642, the king already found himself embroiled in conflict with the Covenanter lords. The parliamentary opposition in England made common cause with James's Scottish foes, promising a 'reformation of religion in the kingdoms of England and Ireland' in exchange for military support. The Scottish army which crossed the border in 1644 was instrumental in destabilizing the royalist position in the North which led to the ultimate defeat of Charles.

By then, though, the leading Covenanter, the Earl of Montrose had defected to the king, a betrayal which failed to save Charles (who was arrested by the Scottish army at Newark in 1646, to be then handed over to parliament, tried and executed three years later). It did, however, serve to spark off a new English intervention when Oliver Cromwell struck north in 1650 in an attempt to snuff out any royalist revival.

That revival did not materialize for the moment; although Charles I's son, the Duke of York (later to become Charles II) was crowned at Scone on New Year's Day 1651, his forces – deprived of the support of many Covenanters – were defeated at Worcester eight months later and he was forced into a nine-year exile on the Continent. The complex interplay of loyalty to the direct descendants of James, nationalist grievance against England and religious rivalry between Presbyterians, Episcopalians and Catholics would continue to dog Scotland and form a potent mix that would find violent expression in the Jacobite rebellions of 1715 and 1745. It was all a far cry from the tranquil, unified land portrayed on Speed's map.

THE KINGDOME OF SCOTLAND

THE DEUCALIDON SEA

The Yles of Hebrides, Called of Pliny Haebudes, of Beda Mevaniae.

THE SCALE OF SCOTISH MILES

PART OF IRELAND

IRISH SEA

PART OF ENGLAND

THE GERMANE SEA

The Yles of Orknay

PLANTATION MAP OF ULSTER, Thomas Raven

The names of the great livery companies of the City of London – the Haberdashers, Vintners, Skinners, Merchant Taylors and Drapers – occupy a great swathe of Ulster between Lough Neagh in the east and Lough Foyle in the west in this 1610 map showing plans for the Ulster Plantation. The project, for the settlement of large numbers of English and Scottish Protestants in the north of Ireland forever changed the demographic balance of the area and lay at the root of much of Ulster's future troubles.

England's uncertain hold on Ireland had been severely shaken by the Nine Years' War of 1594–1603 in which a confederation of Irish chieftains under the Earls of Tyrone and Tyrconnell had come close to expelling the English. The initial peace settlement after their defeat had been comparatively mild: Tyrone, Tyrconnell and the other rebel leaders had been allowed to retain their lands under the terms of English law, rather than Irish feudal custom. Nonetheless, James I, who had just ascended to the English throne, needed to find a way to stop the restive Irish provinces, and especially Ulster which held fast to the Gaelic language and traditions, from revolting again.

The Flight of the Earls in 1607, when Tyrone, Tyrconnell and their confederates fled to Europe to seek Spanish assistance in a new uprising, gave James just the excuse he needed, reinforced by a new rebellion in 1608 by Sir Cahir O'Doherty of Inishowen, which was easily put down. The confiscation of the rebels' land left the English crown holding over half of Fermanagh, most of Tyrone and Donegal, as well as Derry and Cavan. James was now in a position to reward his Scottish followers, who were chafing at their comparative neglect since he had left his homeland for London in 1603.

Plans for the settlement (or 'Plantation') of Ulster by English colonists had been tried before, most notably a scheme promoted by Walter Devereux, Earl of Essex, in the 1570s, but they had been on a far smaller scale and vulnerable to being overwhelmed by the native Irish. Under the influence of the Arthur Chichester, the Lord Deputy of Ireland, James sanctioned a far more ambitious scheme. English-speaking and Protestant colonists were to be recruited and settled on land from which Irish tenants were to be excluded.

Principal landowners or 'Undertakers' were to receive up to 1,200 ha (3,000 acres) and twenty-four able-bodied English or Scottish males were to be settled on each block of 400 ha (1,000 acres). Larger estates were to be banned to avoid the development of independent Ulster magnates. When it became clear the whole project would be ruinously expensive, James tried other expedients. First he established the baronetcy, a kind of hereditary knighthood which could be bought for £1,095 (enough to support thirty soldiers in Ireland for three years). Then, when this was not enough, he turned to the City of London companies to help fund the scheme, asking for an initial grant of £20,000 (and a further £10,000 in 1611) in exchange for which each company would receive a tract of land – 19,800 ha (49,000 acres) for the Skinners, 15,700 ha (38,800 acres) for the Drapers and less for the rest in accordance with their contribution).

By 1610, articles of agreement had been signed with the livery companies – many of who were rather reluctant to commit to such a doubtful enterprise – and the Irish Company was established to supervise the Plantation. Colonists started to arrive as well as hundreds of masons, carpenters and labourers to build the new garrison town of Londonderry. By 1622, a survey established that there were now 6,400 British adult males settled on the plantation lands. Just over half of these were Scottish Presbyterians, whose more radical brand of Protestantism set them even more at variance with their Irish neighbours. By the 1630s these numbers had almost tripled, although thereafter the opening up of fresh colonies in Virginia and New England tended to attract investors and colonists there instead of to Ulster.

Ulster was thrown into turmoil during the 1640s, as the civil war that raged throughout England, Scotland and Ireland embroiled the Plantation. In 1641 the Ulster Catholics staged a major revolt, during which 4,000 Protestant settlers were massacred and 8,000 expelled. An expedition by the Scottish parliament launched in 1642 to aid their Presbyterian brethren led to counter-atrocities against the Irish Catholic population. Only after Cromwell's New Model Army engaged in a brutal campaign of pacification in Ireland in 1650–51 did comparative peace return to Ulster.

A further wave of emigration from Scotland took place in the 1690s, as new colonists came to the Plantation to escape a famine in 1696–8, and by the 1720s Ulster had a Protestant majority. The Plantation had succeeded in establishing a generally pro-English population in Ulster, but in its broader aim of preventing further rebellions in the area it contained the seeds of its own downfall, as the irreparable rupture which it caused in relations between Gaelic and English-speaking communities would further centuries of civil strife, and the eventual British loss of much of Ireland.

The Countie of Antrm

Toome

Dunluic   Colrane

Ferris
Luogh

Lough Beg

Mousnahor

MER   VIN
T

Maherehoy   MAR
CHANT   Ern

CLOTH   NER   S
WOR   AYL   MONGE
ERS   RS   N   E   R   S

KERS   DRAP

HABERDASHERS

Raughryon   SALT
Thomas   ERS
Dungehin

St   DRAP
Phillipps   E
FIS   H   SKIN   NER   S   R   S
Ballehole   Maheresealt

Loughfoyle   MON
GE   Slew Currey
RS

Nuff Culmor   O   Ruff
G   R   Croost
London Derrie   O   L   D

GO
S   M   I   T
H   ES
Nutown

Liffer

Straban.

S. PAULES CHURCH

the Water
house

Quene hythe

Three Cranes

The Eell Schipes

The Gally fuste

THAMESIS

The Bear Gardne

The Globe

PANORAMA OF LONDON,
Claes Jansz Visscher

The sweeping panoramic view of London in 1616 by the Dutch mapmaker Claes Jansz Visscher portrays the city as it was in the reign of James I, providing precious evidence of the appearance of its buildings just fifty years before many of them were destroyed in the Great Fire of London.

Visscher came from a family of Amsterdam mapmakers and the London panorama forms part of a series of such elongated views of major European cities. The four plates used to make it created a printed strip some 2 m (6½ ft) long, showing houses and mansions clustering at the edge of the Thames, from Whitehall in the west to the Tower of London and St Katharine's in the east. By the latter, a 'liberty' which was free from the restrictions of the City guilds, foreign craftsmen clustered, including a community of chart-makers and cartographers. The most prominent building in the panorama is Old St Paul's Cathedral, which would be destroyed by the 1666 fire, and which in Visscher's day had already lost its spire, destroyed by a lightning strike in 1561.

To the south of the river are shown three theatres, the Swan, the Bear Garden and the Globe, the playhouse which William Shakespeare had built in 1599 to house his acting company, the Lord Chamberlain's Players. Shakespeare had already achieved a clamorous fame as the nation's leading playwright with plays such as *Henry V* glorifying the ancestors of the Tudor monarchs. The accession in 1603 of James I, a Scottish Stuart monarch (albeit a great-great grandson of Henry VII), caused a tactful change of emphasis, with plays such as *King Lear*, which harked back to a distant past when Britain was a single realm, an emphasis which would delight the new sovereign, one of whose cherished projects was a formal union between England and Scotland.

Both king and leading subjects had had high hopes of James's reign when he made his way down from Scotland in March 1603 (aided by a loan of £6,000 from the burghers of Edinburgh to pay for the journey). Yet soon resentment at his reliance on Scottish favourites, his tactless treatment of the English parliament, which was far less keen on merging its laws with those of Scotland, and his high-handed insistence on his royal prerogatives (set forth in his 1599 book on the divine right of kings to rule, the *Basilikon Doron* – 'The Royal Gift') soured the mood.

James's new Catholic subjects, too, had hoped for more; his mother, Mary, Queen of Scots had, after all, been brought up a Catholic. But James proved a resolute Protestant, and measures against Catholic recusants, who refused to attend Protestant services, were if anything tightened. Disappointed, Catholic hotheads such as Robert Catesby and Guy Fawkes, a soldier of fortune, began to plot. On the evening of 4/5 November 1605 Fawkes was caught hiding amongst barrels of gunpowder which the conspirators had intended to detonate the next day when King James was due to open a new session of parliament (action which unfolded close to Whitehall Palace at the far left of the Panorama).

Fawkes was arrested and brutally tortured to extract a confession. The other plotters either died in a firefight at Holbeche House, near Dudley in Staffordshire, where royal agents had tracked them down, or were hauled off to the Tower and hung alongside Fawkes on 31 January 1606. The end result was a stiffening of laws against the recusants – the precise opposite of what Fawkes and his band had been trying to achieve – and a temporary thawing of relations between James and parliament (which granted him revenues to the tune of £450,000).

It did not last however, and, exasperated with what he saw as their recalcitrance, James ruled without parliament from 1614 to 1621. Even when he did recall it, his reliance on his new favourite George Villiers, the Duke of Buckingham did nothing to endear the aging autocrat to his leading subjects. When he died in March 1625, arthritic, gout-ridden and wracked with dysentery, James was little mourned. The accession of his impressionable and head-strong son, Charles, however, who had learnt well James's lessons about the theological basis for monarchy and the dispensability of parliament would make his father's reign seem like a golden age.

see more on next page >

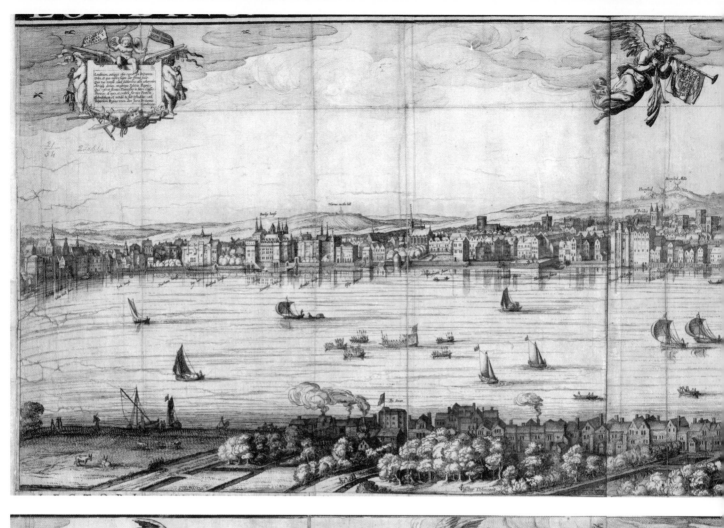

LONDON

FLUVIUS

South **Warke**

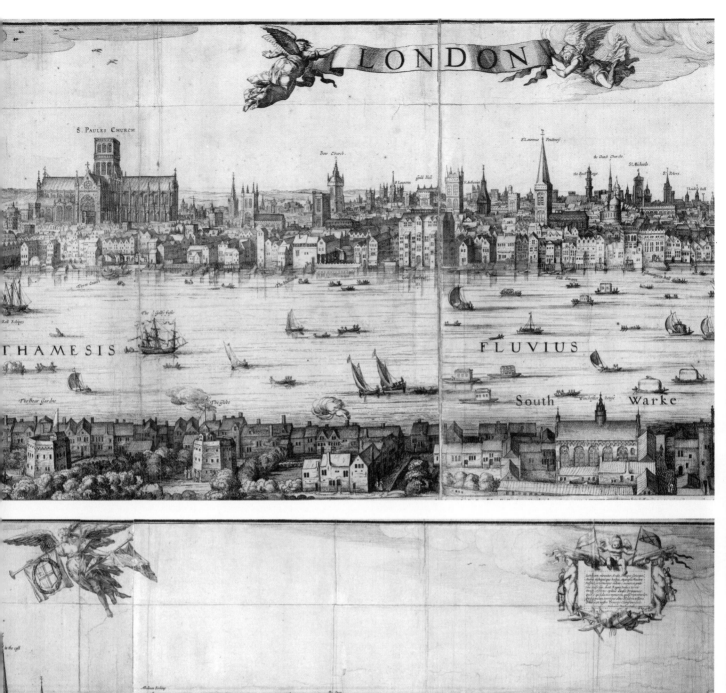

S. PAULES CHURCH   Bow Church   S. Laurens   Gild Hall   S. Anthonie   S. Laurens Poultney   the Dutch Churche   St Michaels   St Peters

LONDON

THAMESIS   FLVVIVS

The Beare Garden   The Globe   South   Winchester house   Warke

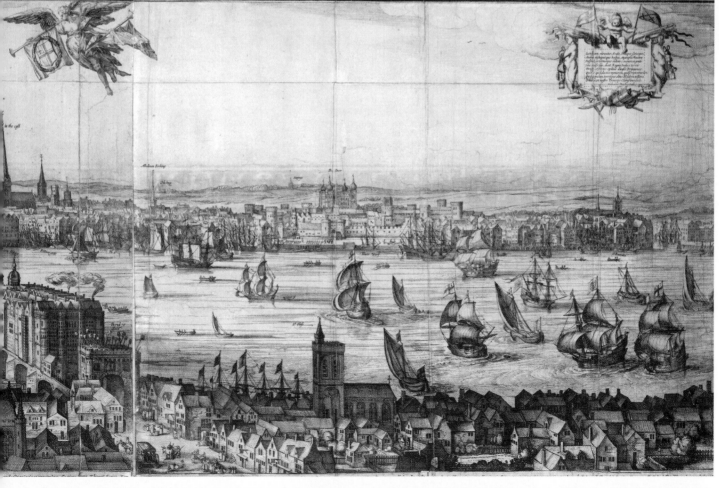

Deuonshire.

	Exceter	Dartmouth	Kings-bridge	Plymouth	Tauestoke	Holdesworthy	Bydiforde	Barstable	Moulton	Bampton	Collombton	Chegforde	Ockington	Hatherley	Chydley	Newtō Bushell	Alhburton	Totnes	Tyuerton	Bradninge	Autre	Hunyton	Ilfarcombe	Culliton	Torrington	Chimley
Axminster NW	20	40	43	50	44	46	42	38	30	20	14	31	35	36	24	28	33	36	18	15	11	7	42	5	39	3
Chimley NW	16	35	35	33	23	17	12	10	8	15	16	15	10	8	20	23	24	29	13	16	21	24	17	28	9	
Torringtō NW	24	40	40	34	23	10	5	8	11	22	15	20	14	8	27	30	29	33	21	25	30	12	14	36		
Culliton E	16	34	39	46	40	43	40	35	29	19	12	20	31	34	20	23	30	33	17	12	7	5	40			
Ilfarcōbe NW	32	53	53	47	36	23	10	7	13	21	30	31	26	20	38	40	40	46	25	30	37	36				
Hunyton NE	12	33	38	44	39	40	35	32	24	15	8	25	28	30	19	24	27	30	12	8	5					
Autre E	9	29	34	39	34	39	34	30	24	16	8	20	22	27	15	18	23	26	12	7						
Bradninge NE	7	30	35	38	30	30	27	23	16	10	3	19	20	21	16	20	23	27	6							
Tyuerton N	11	35	38	40	31	39	23	19	12	6	5	20	20	20	20	24	26	30								
Totnes S	20	8	8	17	18	31	38	40	35	36	29	15	20	26	12	9	6									
Ashburtō SW	16	13	13	17	16	30	33	33	30	31	26	9	16	21	8	6										
Newtō Bushels	13	13	15	22	20	30	34	34	29	29	21	11	18	24	4											
Chidley S	8	16	20	25	21	29	31	30	26	25	17	10	16	20												
Hatherley NW	19	34	21	26	16	10	12	14	15	23	13	12	6													
Ockingtō W	16	28	26	22	13	13	17	19	18	24	22	7														
Chegford SW	12	21	21	20	14	20	24	24	21	24	20															
Collōbton NE	9	33	37	40	34	33	27	23	16	9																
Bampton N	18	40	43	45	36	30	23	17	10																	
Moulton NW	19	41	42	40	30	20	13	8																		
Barstable NW	25	45	48	40	30	18	7																			
Bydiforde NW	27	45	44	37	27	12																				
Hodesworthy NW	29	39	36	27	17																					
Tauestoke SW	25	25	20	10																						
Plymouth SW	32	22	15																							
Kings-bridg SW	28	8																								
Dartmouth S	24																									

The seauerē sea N.

Cornwall West

ye Brittaie sea S.

13

A DIRECTION FOR THE ENGLISH TRAVILLER,

Matthew Simons

A Direction for the English Traviller by Matthew Simons, from 1635, sought to resolve a serious difficulty for those traversing the realm. There was simply a shortage of accurate information about the distances which needed to be travelled between places, making accurate route-planning next to impossible.

There had long been itineraries (as far back as Matthew Paris in the thirteenth century in the case of England, and even longer ago if Roman imperial itineraries are included) which gave rough directions from point to point along a route and, on occasion, some indication of the time a traveller should expect to take between stages, but these were crude tools. In 1625 John Norden produced a distance table showing the number of miles between major towns and Matthew Simons' maps provided a refinement of this, adding thumbnail sketches of each county so that travellers could more easily visualize the routes they were calculating.

Stuart travellers still faced a number of very real practical difficulties. Not least among them was that manuals such as Simons's included only small maps – if they had any at all. The larger road atlases were unwieldy and hardly practical items to carry around on horseback. Only with the publication of Emanuel Bowen's *Britannia Depicta* in 1720 was there a really usable portable road atlas for travellers.

The state of the roads was a perennial difficulty. Local parishes were generally responsible for their upkeep but many did not have the resources, or chose not to direct them to the maintenance of the road system. In 1523 John Fitzherbert had noted in his *Book of Husbandry* that the roads around London, being built on chalk, had a terrible tendency to subside, but nothing was done. Thomas Procter did call in 1607 for the building of entirely new roads constructed on timber lattices, but he was roundly ignored.

Instead Tudor and Stuart governments issued plaintive appeals for the better maintenance of the roads and adopted a piecemeal approach to urgent repairs. In 1555 a more general statute made provisions for seven years that all parish constables and churchwardens were to meet annually and elect surveyors who would attend to the state of the roads, but the labour levy they were allowed to call upon of four or six days per man per year was inadequate to maintain, let alone improve, more heavily trafficked roads. In a desperate attempt to minimize the damage to roads caused by overloading, in 1663 an Act was passed allowing surveyors to levy tolls depending on the weight of vehicles (at 1d for those on horseback up to 12d for waggons).

If a traveller did not get bogged down in ruts caused by the overuse of inadequate roads by heavy carts or disappear into a sink-hole where the road surface simply collapsed, there was the problem of wayfinding. Knowing the distance in miles was not always as useful as it seemed, for not all miles were equal: 30 Scots miles in 1618 were equivalent to 40 English (or statute) miles, leaving English travellers unfamiliar with the discrepancy weary and miles from their hoped-for destinations. There were few signposts to guide those who became lost – only in 1696 was a law passed requiring surveyors to install them indicating the distance to the nearest market town – so that once a traveller lost his way, resort to local guides (and the resulting expense) was often the only recourse.

Although the roads were in truth not as bad as contemporary writers often portrayed them and most travellers did reach their destination unmolested by thieves or vagabonds, travel was not fast. The 737 km (458 miles) that Sir Robert Carey travelled in three days from London to Edinburgh carrying news to James VI at Holyroodhouse that he was now king of England was an exceptional case, made possible by relays of post horses belonging to the government. For private travellers these cost up to a penny a mile in 1625 and so were an occasional luxury. Even those owning their own horses, or able to hire them, rarely covered more than 32 km (20 miles) a day, and in lumbering carts or even slightly more luxurious coaches the going was much slower.

For all those travelling the Stuart road network, knowing the distance, therefore, significantly mitigated the uncertainties of travel. The waggoners, carriers, porters, pedlars, soldiers, entertainers, messengers and merchants would all benefit from the intelligence provided by Matthew Simons' work.

MAP OF THE START OF THE ENGLISH CIVIL WAR,

Wenceslaus Hollar

This dramatic map of the British Isles, criss-crossed by armies and surrounded by fleets, was drawn by the Bohemian cartographer Wenceslaus Hollar, as a way of explaining the origins of the English Civil War, but also as a warning to the British, as it compares its genesis to that of the 'Thirty Years' War which had torn apart Hollar's homeland over the preceding two decades (the battle scene is of an encounter near Prague, for example).

In 1638, England had been at peace. By 1649 the King had been beheaded and a republic installed, whose increasingly radical aspect caused the army to step in and install a type of military dictatorship under Oliver Cromwell, as Lord Protector, in 1653. The root cause of this slide into disaster was the king's need for money, both to fund the extravagance of the court and to conduct a war in Scotland, needed after the outbreak of rebellion there in 1638 (a moment marked by Hollar as A on the main map).

Charles had been schooled by his father James I on the divine right of kings to rule and resented any constraints on his freedom of action. His attempt to raise taxes which did not need the consent of parliament, such as Ship Money, a medieval relic which the king reinterpreted to levy heavy taxes on English ports, caused outrage. Short of funds, however, Charles was forced to summon parliament, the only body able to vote him substantial taxes. The 'Short Parliament', which assembled in April 1640, proved recalcitrant. Under the leadership of John Pym, it took the opportunity to present the king with a long list of grievances and to try to oppose further military operations against Scotland. Charles, angered at such treatment dissolved it (depicted by Hollar in picture F).

Charles's position deteriorated as the Scots advanced into northern England in summer 1640 and he was forced to pay them a subsidy to stop them moving any further south. In November 1640 he had to recall parliament (whose long session, until 1648 gave it the nickname the 'Long Parliament'). In an effort to win the backing of parliamentary moderates, Charles sacrificed his favourite, the Earl of Strafford, who was accused of treason by parliament and subjected to Attainder (a process

see more on next page >

by which parliament simply voted on his guilt) and executed on 12 May.

The demands did not stop. On 10 May parliament passed a bill depriving the king of the right to dissolve it. In November it approved the Grand Remonstrance, a long list of grievances against the king and the following month it asserted control over the army. Alarmed, Charles resolved to put a stop to parliament's impudence, and on 4 January 1642 did what no monarch had done before him. He entered the House of Commons accompanied by armed guards to arrest the five MPs (including John Pym) he regarded as the ringleaders of the movement against him (the moment is shown by Hollar in picture I).

The move failed as the MPs had been forewarned. On seeing their absence, the king curtly commented 'It seems the birds have flown', but before long he, too, was forced to flee the capital. By now parliament had approved the Militia Ordinance establishing local Trained Bands which did not answer to the crown. Charles began to muster his own supporters and on 22 August 1642 the king raised his standard at Nottingham. The Civil War had begun. Although the first major action (at Edgehill) did not take place until October, the complex series of campaigns which broke out (involving Scotland and Ireland, as well as England) meant the country knew little peace until 1648, when the king's last army (in alliance with the Scots) was destroyed and he was executed the following year.

All that was in the future, but Hollar's map contains a vivid sense of the destructive nature of civil war. Many of its images deal with the Thirty Year's War, which had erupted in 1618 and turned Europe into a slaughterhouse as Protestant and Catholic armies fought and pillaged across Bohemia, Austria, Germany and France. The admonition that the suppression of political dissent might lead to war is encapsulated in the large image to the right of the main map, which shows the Battle of the White Mountain, fought in Bohemia in 1620, at which the Protestant forces of the Elector Frederick V of Bavaria were crushed by those of the Catholic Emperor Ferdinand II. Twenty-eight years of relentless conflict followed until peace was finally brokered at Westphalia in 1648. Hollar's warning was ignored and Britain did not achieve a final settlement for eighteen years, until the Restoration of 1660, when Charles II assumed his executed father's throne.

Sed nulla potentia longa est.

Quo non discordia Cives,

SIEGE OF NEWARK,
Richard Clampe

A dense line of earthworks, fortifications and redoubts encircle the town of Newark in this 1646 plan of the siege of the town by parliamentary forces during the English Civil War. It was one of the Royalists' last strongholds and its fall, choked by the steady tightening of the siege, marked almost the end of Charles I's hopes of keeping his throne.

The plan was drawn up by Richard Clampe, an engineer working for parliament, towards the end of the siege. With the keen eye of a military draughtsman, he sketched out the 'Line of Circumvallation', a system of earthworks punctuated by strongpoints which had steadily encircled the town, and the star-shaped redoubts which the besieging forces constructed: to the west – at the bottom edge of this map - 'London', where parliament's troops (some 7,000 strong) under Colonel-General Poyntz were based; and to the east 'Edinburgh', the headquarters of the Scottish contingent led by the Earl of Leven.

The siege, which began in November 1645, was the third that the Nottinghamshire town had endured in as many years. The trials which is inhabitants suffered – around a third would die before the final siege was over – were a result of their loyalty to Charles I which left Newark as one of his last strongholds when the tide of war turned against him. The conflict had broken out in 1642 after years of tension between king and parliament over the right to approve taxation (which parliament claimed as its domain, and Charles doggedly claimed to be a royal prerogative). It was war with Scotland that proved Charles's undoing. The funds needed to put down a rebellion which broke out there in 1638 forced him to recall parliament to vote new taxes to pay for an army. The parliamentary side, led by John Pym, sought to force the king into concessions which he simply would not grant, and by October 1642, debate had been replaced by the clash of arms as the first pitched battle of the war was fought at Edgehill. Initially the war went well for the king, but the intervention of the Scots in 1644 and the reorganization of the parliamentary forces into the New Model Army the following year tipped the balance against the royalists, who steadily lost ground, until Charles was confined to a narrow area of land between Oxford, his headquarters, and Newark.

A parliamentary siege of Newark had been relieved in March 1644 by Prince Rupert of the Rhine, Charles's charismatic German nephew, and cannon captured were then used to strengthen the town's defences. But royalist raids from there into the Midlands during the summer of 1645 proved too much of a nuisance for parliament, who despatched a contingent to capture it once and for all. Hearing of the approach of the parliamentarians, the town's governor Lord Belasyse ordered the digging of further earthworks, but by 4 November Poyntz's men had arrived and the trap began to close. The Scots reached Newark three weeks later, released by the destruction of the Scottish royalist army at Philiphaugh, near Selkirk, in mid-September.

Belasyse's defence proved surprisingly stubborn and although the circle of siege-works around Newark was complete in March, he would not surrender. Poyntz even tried to dam the River Trent to silence the royalist corn-mills and starve out the town. In the end, Newark was betrayed by the king. Secret negotiations had been underway for some time through an envoy of Cardinal Mazarin of France, aimed at forging a separate peace with the Scots (whom Charles hoped would then somehow come to his aid against parliament). On 27 April, as the New Model Army grew dangerously close to Oxford, the king fled in disguise and made his way to the Scots camp at Newark. There he surrendered himself to the new Scottish commander, Lieutenant-General Leslie. Part of the price for his safe-conduct was the giving of the order to surrender Newark. When he received it, Belasyse is said to have wept, but he felt he had no choice but to obey his sovereign. On 6 May 1646 he marched out of the ruined town with his surviving troops.

The manoeuvre did Charles little good. Relations between parliament and the Scottish Covenanters had become fraught, but in January 1647 the king was handed over to parliament as part of a deal to improve them (smoothed by a hefty payment to the Scots). Even in captivity, Charles carried on plotting, engineering a new alliance with Scotland in December 1647. This time the price was the establishment of Scottish-style Presbyterianism in England, but the king's complex scheme unravelled when the Scots army was defeated at Preston in August 1648. Friendless and still deluded that he could somehow argue or command his way out of trouble, Charles was tried by a parliamentary commission, sentenced, and on 30 January 1649 was beheaded outside the Banqueting House of Whitehall Palace.

His last hope, though, had in truth died long before, a suffocation of the royalist spirit that took place in the cheerless, desolate fight around Newark. Once he surrendered himself to the Scots there, Charles became, not a king but a pawn, a piece in a complex game that he no longer had any power to control.

the Royal Cittadel.

Mount Batten

St Nicolas Iseland

the poole or Harbor

The Towne of Pleymouth

the Royal Cittadel.

Fishers Nose

the Lower gate

Bath Bastii

Governors house

Great Storehouse

ponder house

Ruperts Bastii

The Royal
Cittadell

Charles Bastii

Corps de Gaurd house

James
Bastii

Caterins
Bastii

The place of the
ould Castle of the
Towne.

Smarts Key

Barbican

The Poole or Harbor

Chart of the Town of Pleymouth

English feete or 200 Jacres

PLAN OF PLYMOUTH,
Bernard de Gomme

This 1666 plan of Plymouth by Bernard de Gomme (accompanied by a panorama) shows his design for restoring and improving the strength of the town's defences, which had been severely damaged during an unsuccessful three-year siege by Royalist forces during the English Civil War. At the time of the siege, de Gomme was the Royalists' chief engineer, and he was then called back into service after the Restoration to redesign Plymouth's fortifications, having had some responsibility for their damage in the first place.

Plymouth was an ancient town. Its name first appears (as Plimmuth) in 1211, but previously known as Sutton, it had long been a port, given the impetus for growth by the discovery of tin on West Dartmoor around 1150. From then, for centuries, tin, fish, hides, lead, wool and cloth left from Plymouth, enriching the merchants who imported wine, iron, fruit and wheat in return. As a rich south-coast port, the town inevitably attracted the attention of French raiders; in 1403 a large force of thirty ships and 1,200 men under the Sieur du Chastel landed, stormed the walls and briefly occupied a section of the town (which was forever after known as the 'Bretonside').

The scare led Henry V to order the building of the first castle at Plymouth in 1416, fortifications which were progressively improved during the reigns of Henry VIII and Elizabeth I, when attacks from Spain became a more pressing cause of concern. It was from Plymouth that Sir Francis Drake sailed in 1587 to 'singe the king of Spain's beard' by attacking Cádiz and from where he set out to act as vice-admiral in the English fleet to oppose the Spanish Armada in 1588.

By the seventeenth century Plymouth had become a Puritan town, and sentiment was strongly in favour of Parliament in its quarrel with Charles I. When the Civil War broke out in August 1642, most local notables declared for Parliament (though there were exceptions, such as Edward Hyde, MP for Saltash, whose loyalty to the crown was rewarded with the Earldom of Clarendon). When Charles raised his standard at Nottingham in August 1642, he made the mistake of summoning Plymouth's governor, Sir Jacob Astley, to be his major-general of foot, opening the way for the garrison to be taken over by Parliamentary sympathizers.

Under the command of Colonel Ruthven, Plymouth weathered the first Royalist attack in November 1642 under Sir Ralph Hopton, but after a counter-attack into Cornwall failed, the Parliamentary forces found themselves besieged inside the town, a siege that on and off lasted for the next three years. Earthworks were thrown up, ditches dug and ramparts reinforced. Reinforcements were shipped in by sea until the garrison reached a strength of 9,000 troops. There was a crisis in August 1643 when a new Royalist force under Prince Maurice advanced, capturing outlying forts but failing in two concerted attempts to storm the walls. Finally, Maurice pulled back, hoping to starve Plymouth into submission, but just at that moment an enormous shoal of pilchard providentially swam into the harbour, providing the grateful townspeople with an unexpected bounty to survive the winter.

In September 1644, the king himself menaced Plymouth with an army of 15,000 men and twenty-eight guns, but he, too, failed to take the town. Only the general collapse of the Royalist position in the west later the following year (after the catastrophic defeat at Naseby in June 1645) undermined the Royalist siege of Plymouth. On 12 January 1646, the Royalists abandoned the siege and two months later Oliver Cromwell led a triumphant procession into Plymouth to celebrate its successful defence. The siege had been long, costly and damaging – the town's Committee of Defence had been forced to borrow up to £5,000 a month to fund its defence – the fortifications had been pummelled by Royalist artillery and there were only five ships left in harbour.

After the Royalist defeat, de Gomme left the country in 1646, his reward finally coming with the Restoration in 1660, when he was appointed Surveyor-General of Fortifications (and promoted to Surveyor-General of Ordinance in 1682). In 1662 the Navy Board had ordered the building of a new dockyard at Plymouth, and so the need to restore its defences had become urgent. De Gomme sketched out a new star-shaped fort in the manner of the great French military architect, Vauban, incorporating parts of the old fort. The resulting structure, mounting over 150 guns, was almost impregnable. Not everyone was impressed, of course, and Samuel Pepys, the great diarist and a naval administrator himself, acidly remarked that 'De Gomme hath built very silily'.

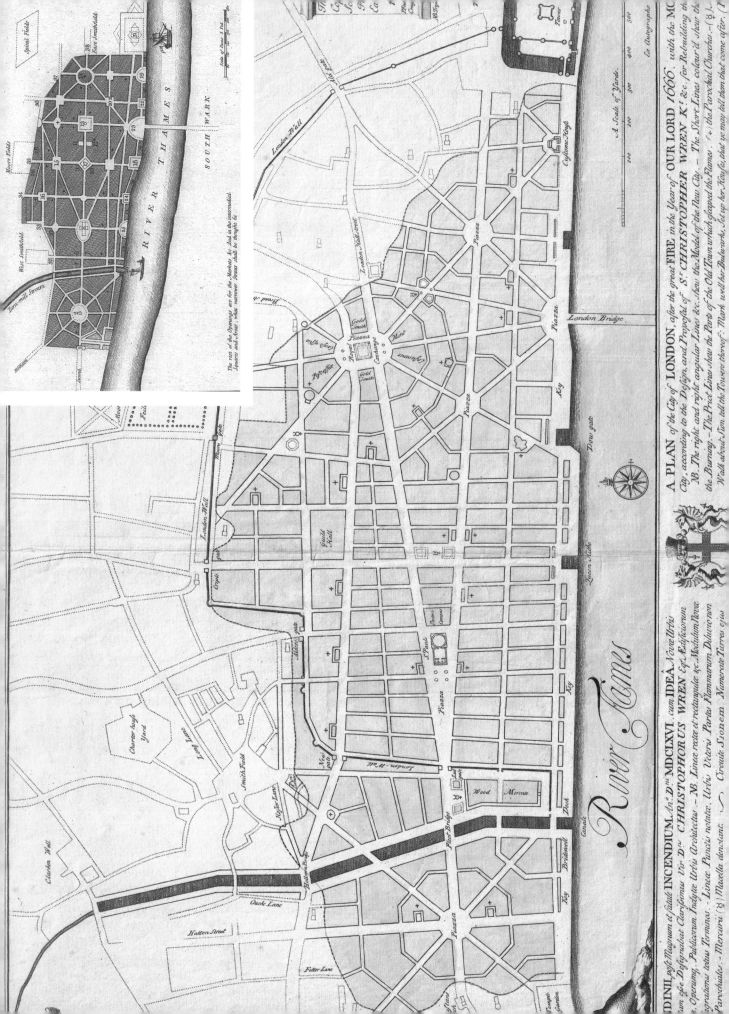

A PLAN of the City of LONDON, after the great FIRE in the Year of OUR LORD 1666, with the MO...
... City, according to the Design, and Proposal of Sʳ CHRISTOPHER WREN Kᵗ &c. for Rebuilding the
... Designabat Clarissimus Vir Dⁿᵘ CHRISTOPHORUS WREN Esqʳ, Ædificiorum
... NB. The right and right angular Lines &c. shew the Model of the New City. — The Short Lines colour'd shew the
... Openings, Include Urbis Architectu. — NB. Lineæ rectæ et rectangulæ &c. Modulum Novæ
... the Burning. — The Pict Lines shew the Parts of the Old Town which escaped the Flames. + the Parochial Churches — (§.)
... gratissimi totius Terminus. — Lineæ Puncti notatæ. Urbis Veteris Partes Flammarum Diluvio non
... Walk about Sion tell the Towers thereof: Mark well her Bulwarks, Set up her Houses, that ye may tell them that come after. (P...
... Circuite Sionem Numerate Turres ejus
... (§) Macella denotant:— Mercurii [?]

Spital Fields

East Smithfield

Moor Fields

West Smithfield

Turn-mill Stream

RIVER THAMES

SOUTH WARK

The rest of the Openings are for the Markets &c. And in the intermediate
Squares and Areas: what narrower Streets shall be thought fit

Scale of Paces 5 Fab.

London Bridge

A Scale of Yards.

London Wall

London-Wall-Street

Custome House

Piazza

Piazza

Piazza

Piazza

Piazza

Gold Smiths

Royal Exchange

Piazza

Mint

Post office

Gold Smiths

Key

Dew gate

Moor Field

Charter-house Yard

London-Wall

Guild Hall

Clerken Well

Long Lane

Smith-Field

Fleet Bridge

Bridewell

Bridewell Brid.

Oxeole Lane

Hatton Street

Fetter Lane

Temple Garden

Kayes Lane

Alder.s gate

New gate

London M.M.

Wood

Mercate

Canale

Dock

Key

S Paule

St Paule

Piazza

Piazza

Key

Lucas Siche

Cripple gate

Cripple

River Tames

GREAT FIRE OF LONDON MAPS,
Christopher Wren and John Evelyn

They are visions of a London that never came to pass, of a phoenix that never arose from the ashes of a city devastated by the Great Fire of 1666. The plans submitted by the ambitious young architect Christopher Wren (main map) and the diarist John Evelyn (inset) envisaged neat orderly squares and wide vistas to replace the tight maze of streets through which the fire had blazed. In the end, though, pragmatism trumped their imaginary tours de force and their schemes were never carried out.

The Fire had broken out shortly after midnight on 2 September 1666 in a baker's shop in Pudding Lane, close to the city's heart. The streets around were a tinder-box of timber-framed houses and merchants' warehouses that provided ample combustible material for the fire to spread. There was no formal fire-fighting service and arguments over who would pay compensation to the owners of buildings whose premises were pulled down to retard the spread of the flames wasted precious hours as sparks leaped from house to house, spreading the blaze to new districts.

By Monday 3 September the fire had consumed a large part of the southern section of the city, spreading out along the riverfront as far as Blackfriars. Yet there was worse to come. The next day St Paul's Cathedral caught fire and the lead of its roof melted, dripping molten fire onto the heaps of volumes that the booksellers, who plied their trade nearby, had stored for safekeeping inside the building. By the time the wind dropped and the spread of the fire began to slow, it was clear the city had suffered catastrophic damage. Some 160 ha (395 acres) had been burnt, including 87 parish churches and over 13,000 houses. Around 200,000 people had fled their homes, many of them encamped in Moorfields, Highgate and Islington.

It was vital to see to the needs of the displaced, to avoid their anger turning into an uprising and the king soon gave orders for them to be provided with tents and food. Rebuilding took longer, as even six months later the debris still smouldered in places, making clearing impossible. Some preliminary measures were possible and on 13 September an order was issued that all new buildings had to be of brick or stone and that all new streets should be wide enough to allow vehicles to pass each other (and also as a side-effect making it harder for flames to leap from one side of the road to the other). Wooden doors and door-frames were also required to be set back from the wall, again reducing the opportunities for fires to spread (and lending a certain uniformity to the newly rebuilt quarters).

Proposals soon flooded in for the reshaping of the City. Among them was one from Robert Hooke, the scientist and leading light in the Royal Society. More practical plans were submitted by Evelyn, who believed that all the main public buildings should be re-sited to the edge of the city along a wide embankment. Wren's plan was more grandiose, calling for wide avenues to radiate from St Paul's and the Royal Exchange and the construction of an enlarged embankment along the Thames.

Wren's plan was too radical and city guilds and private merchants were keen to begin rebuilding on their own terms and on the site of their old premises. By 1670 most of them had done so and Wren's scheme was shelved for good. He did, however, win himself a vital role in London's reconstruction as the architect of most of the churches built to replace those lost in the Great Fire. Most notable of these was St Paul's, with its great dome, on which he laboured for thirty-five years after work began in 1675. Wren rebuilt another fifty-one city churches, but St Paul's was his masterpiece. Far more than his visionary plan for London, it was his monument, a notion reinforced by the epitaph on Wren's tomb inside his great creation: *Lector, Si Monumentum Requiris, Circumpsice* ('Reader, if you require a monument, look around').

CHART OF THE BRISTOL CHANNEL, John Seller

John Seller's 1671 chart of the Bristol Channel documents the shoals and currents, the shallows and shipping channels in one of Britain's most strategic waterways. The chart formed part of the first attempt at a maritime atlas of Britain at a time when concern over the security of the country's coastline was growing as a result of several wars with a new seagoing superpower, the Dutch.

London-born Seller was originally a compass-maker, whose career nearly came to an abrupt end in 1662, when he was implicated in an alleged plot against Charles II. The conspirators, led by a certain Thomas Tonge, planned to kidnap and kill the king and replace him with his pro-Catholic brother, the Duke of York. Seller was arrested after the plot was unmasked and he was accused of having passed on information about the purchase of arms by Tonge and his confederates. Although he was found guilty of treason, for some reason Seller was the only one of the five accused who was not executed.

Incarcerated in Newgate prison, Seller wrote heartfelt pleas to the Duke of York for a pardon and was eventually released. He used the chance of his reprieve to change the direction of his career, beginning to publish maps and to compile his own. In 1671 he produced *The English Pilot*, which for the first time included detailed maps of the waters off Britain's coasts, including the Bristol Channel plate. Critics, though, accused him of plagiarizing Dutch maritime atlases (known as 'Waggoners' after the pioneering Dutch naval cartographer Lucas Janszoon Waghenaer). Seller's change of fortune was confirmed when, in the same year he completed *The English Pilot*, he was appointed Hydrographer in Ordinary to the king.

Perhaps Charles II was grateful not to have to rely so heavily on Dutch charts. A first Anglo-Dutch War had already broken out under the Commonwealth in 1652. The Dutch had gained markets during the English Civil War, when British trade was disrupted, and the passing of a Navigation Act in 1651 which forbade the carriage of goods to English possessions overseas in any but English ships caused Anglo-Dutch relations to deteriorate.

The two-year war was inconclusive, although the Dutch lost hundreds of ships to English privateers in the Caribbean, the treaty that ended it changed little. Resentment at perceived Dutch encroachment into British markets continued to fester: as illustrious a person as General George Monck (who had brokered Charles II's restoration in 1660) remarked that 'What we want is more of the trade the Dutch have now', and when Dutch ships began to seize vessels of the Royal Africa Company off the coast of Guinea, it was taken as a pretext for war. Although the English navy enjoyed some success, defeating the Dutch at Lowestoft in a titanic clash that saw 100 ships fighting on each side, they also suffered the worst defeat in their naval history. On 10 June 1667, a large Dutch flotilla under Admiral de Witt evaded the English patrols and crept up the Medway, seized the fort at Sheerness and, having broken a chain that prevented vessels sailing any further, fell upon the English fleet at anchor in Chatham. Thirteen vessels were burnt, including three large warships, and the English flagship, the *Royal Charles*, and one other ship were captured and towed back to Holland.

It was a humiliation and support for the war flagged, leading to an early peace under which the Navigation Act was amended to be less unfavourable to the Dutch. The main result as far as England was concerned was the occupation in 1664 of New Amsterdam, the principal Dutch colony in North America, later to be much better known as New York. Another war broke out in 1672, in part because of Charles II's determination to forge a French alliance, which in turn meant joining Louis XIV's campaign against the Dutch. The naval side of the war was characterized by two inconclusive attempts to destroy the Dutch fleet at Schooneveld, whose failure saved the Dutch Republic from reverses they had suffered at the hands of Louis' ground invasion. The Second Treaty of Westminster which brought an end to the war in 1674 formalized British possession of New York (and reciprocated by recognizing the Dutch occupation of Suriname in South America). Perhaps more importantly it led to the marriage of Charles's niece Mary to the Dutch Prince William of Orange, a match which would ultimately lead to his accession to the English throne as William III in 1688.

Amid this atmosphere of trading, diplomatic and military rivalry with the Dutch, any advantage the English navy might acquire was eagerly sought. Seller's maritime charts, therefore, were a most welcome boon, and may help explain his rapid change of fortune from condemned criminal to feted royal official.

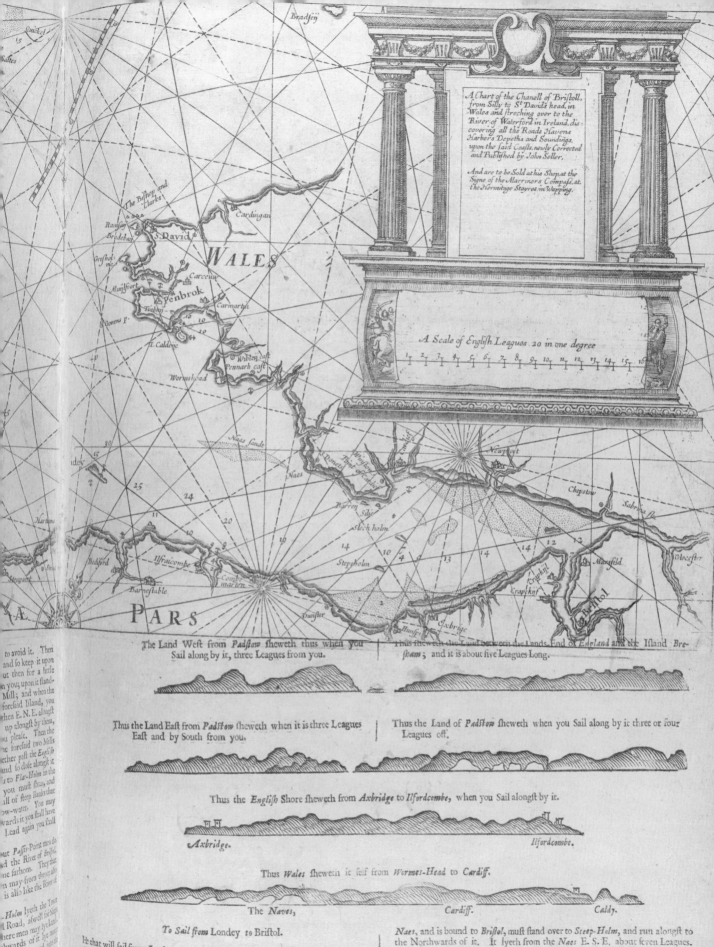

A Chart of the Chanell of Bristoll, from Silly to St Davids head, in Wales and streching over to the River of Waterford in Ireland, discovering all the Roads Havens Harbors Depeths and Soundings, upon the said Coasts, newly Corrected and Published by John Seller.

And are to be Sold at his Shop, at the Signe of the Marrinor & Compass, at the Hermitage Stayres in Wapping.

A Scale of English Leagues. 20 in one degree

WALES

PARS

Æ

The Land West from *Padstow* sheweth thus when you Sail along by it, three Leagues from you.

Thus sheweth the Land between the Lands End of *England* and the Island *Bresham*; and it is about five Leagues Long.

Thus the Land East from *Padstow* sheweth when it is three Leagues East and by South from you.

Thus the Land of *Padstow* sheweth when you Sail along by it three or four Leagues off.

Thus the *English* Shore sheweth from *Axbridge* to *Ilfordcombe*, when you Sail alongst by it.

Axbridge.   *Ilfordcombe.*

Thus *Wales* sheweth it self from *Wormes-Head* to *Cardiff*.

The *Naves*,   *Cardiff*.   *Caldy*.

To Sail from Londey to Bristol.

He that will sail from *Londey* to *Bristol*, must run alongst the English Coast until he come within the Point of the *Naes*, for to avoid the *Naes* Sand. And then through between the *Holms*, leaving the *Steep-Holme* on Starboard and *Flat-Holm* on Larbord side, men may also with Ships of small draughts sail about to the Southwards of *Steep-Holm*, but it is there so Shoally that there remaineth at Low-water no more than two fathom water. Under *Steep-Hol..* men may anchor where they will in four...

Naes, and is bound to *Bristol*, must stand over to *Steep-Holm*, and run alongst to the Northwards of it. It lyeth from the *Naes* E. S. E. about seven Leagues. A little to the Westwards of *Milford Haven* lyeth two little Islands, the Southermost is the smallest, called *Stockholm*, and the Northermost *Scaline*; about two Leagues N. N. W. and N. W. by N. From thence lyeth the Island *Ransey*; and betwixt these Islands the Land hath a great Bay, called the *Broad-Bay*. *Ransey* lyeth at the North-Point, and *Scaline* at the South-Point of the Bay; therein alongst by the Shore is good riding for North North East, East and...

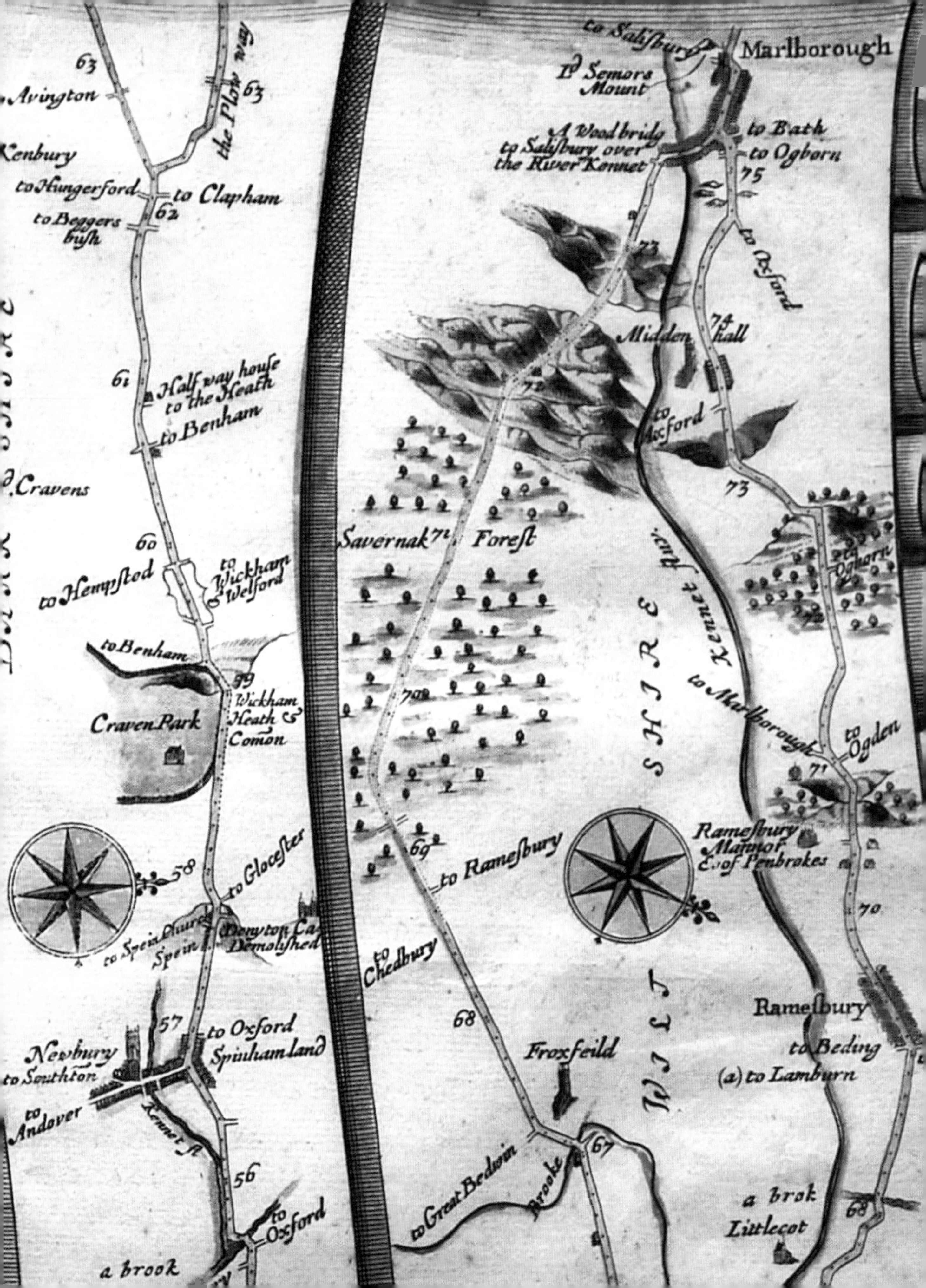

63
Avington

Kenbury
to Hungerford
to Beggers bush
to Clapham
62

the Plow way
63

to Salisbury
Marlborough
L.ᵈ Semors Mount

A Wood bridge
to Salisbury over the River Kennet
to Bath
to Ogborn
75

73

61
Half way house to the Heath
to Benham

d. Cravens

60
to Hempsted
to Wickham Welford

to Benham

Craven Park

Wickham Heath & Comon
59

Savernak 71 Forest

Midden Hall
74

to Aicford
72

to Aicford

73

to Ogborn
72

70

to Glocester
to Ramesbury
58

to Spein Church Spein
Newton Ca: Demolished

to Chedbury

69

to Ramesbury

to Marlborough
Ogborn

to Ogden
71

Ramesbury Mannor E. of Penbrokes
70

57
to Oxford
Spinhamland

Newbury
to Southton

to Andover

Kennet R.

56

a brook

to Oxford

68

Froxfeild

to Great Bedwin
Brooke
67

Ramesbury
to Beding
(a) to Lamburn

a brook
Littlecot
68

WILTSHIRE
BERKSHIRE
Kennet River

BRITANNIA, John Ogilby

The post-roads from London to Bristol and the distances between towns and villages on the way are carefully catalogued in this 1675 extract from John Ogilby's *Britannia*, the very first road atlas of Britain. Its 100 strip maps provided for the very first time a real guide to the vicissitudes of travel between the country's major towns.

Scottish-born Ogilby had had a chequered career before his appointment as 'His Majesty's Cosmographer and Geographer' in 1674 with a brief to produce a new survey of his royal master's realm. As a youth his family had fallen on hard times and he had found employment as a dancing master and choreographer until an accident left him with a limp. He used the contacts he had made to secure a position with the Earl of Strafford as tutor to his children in Ireland, and while there founded the Theatre Royal in Dublin. Yet disaster struck; Strafford was disgraced (and in 1642, executed) and the theatre went bankrupt, leaving Ogilby penniless.

It was not until the Restoration in 1660 that his fortunes began to turn. England had suffered a process of political decay since the heady days of the Civil War when radical parliamentarians had the king executed in 1649 and founded their own republic, a purer political form, or so they considered. Instead, radicalism spiralled out of control, and to rein it in a more hum-drum form of military dictatorship emerged under the army commander, Oliver Cromwell, who was appointed Lord Protector in 1653.

Cromwell's rule provided a steady, if surprisingly uninspiring, hand, but by the time of his death in September 1658, the country yearned for change. The nomination of his brother Richard as the new Lord Protector made matters worse. Entirely unsuited to the role – he acquired the unflattering nickname 'Tumbledown Dick', Richard did little save oppose the dissolution of the Third Protectorate Parliament, which was supportive of him, and when the army stepped in again and forced the re-convocation of the old Rump Parliament in May 1659, he was rudely sidelined.

England now had no proper leadership and a slide to anarchy beckoned. Royalist sentiment was waxing and to appease it General Monck, the new army commander, sent word over to The Hague, where the exiled Charles Stuart was biding his time, and invited him to resume the crown as Charles II. The young prince had had almost as varied a time of it as Ogilby.

After his father's death, he had fled to Scotland, but his Scots allies were defeated at Dunbar in 1650 and then, when they were prevailed upon to invade England, were again soundly trounced at Worcester (after which Charles was said to have escaped detection by hiding up an oak tree).

Charles then fled with a few loyal servants and took the road to Bristol. The going was slow as he feared betrayal and lacked the information that a guide such as Ogilby's might have provided. Finally, after wanderings which took him to Bridport and then Brighton, always fearing someone might take the £1,000 reward for turning him in, he took ship for France and eight years of exile. When finally he returned to England, he took advantage of all the pleasures of royalty that he had been denied. There was an outburst of hedonism: theatres were reopened, gaming dens restored and the king indulged in a string of mistresses including the actress Nell Gwynne and Barbara Villiers, Countess Castlemaine.

At a political level, Charles sought better relations with France (and, against parliament's bitter opposition, toleration for Catholics). It was this which brought him in 1670 to sign the secret Treaty of Dover with Louis XIV (mediated by his sister Minette who was married to the French king's brother). In it Charles promised to join France in a war with the Netherlands and, ultimately to convert to Catholicism (a promise he never carried out).

It was perhaps this, and a desire to assert mastery over his country, in a way in which an activist parliament seemed determined to deny him, that led Charles to commission Ogilby to produce his maps from 1671. It took three years, and 42,000 km (26,000 miles) of travel by the determined surveyor – whose salary was a meagre £13 6s 8d a year, but whose expenses amounted to over £20,000 – and when it was finished the result came to 100 pages of strip maps, weighing over 15 pounds. He died in 1676 and two further planned volumes were never produced. It was not exactly a practical guide (a portable version was only produced in 1719), but it would help overcome the vagaries of travel which had once led Samuel Pepys and his wife to pay £1 2s 6d for a guide for the short distance from Huntingdon to Oxford. And for Charles, it might serve as a military manual should he ever need French help to reimpose Catholicism and as a salutary reminder of those dark days when he himself was a hunted fugitive on the road to Bristol.

see more on next page >

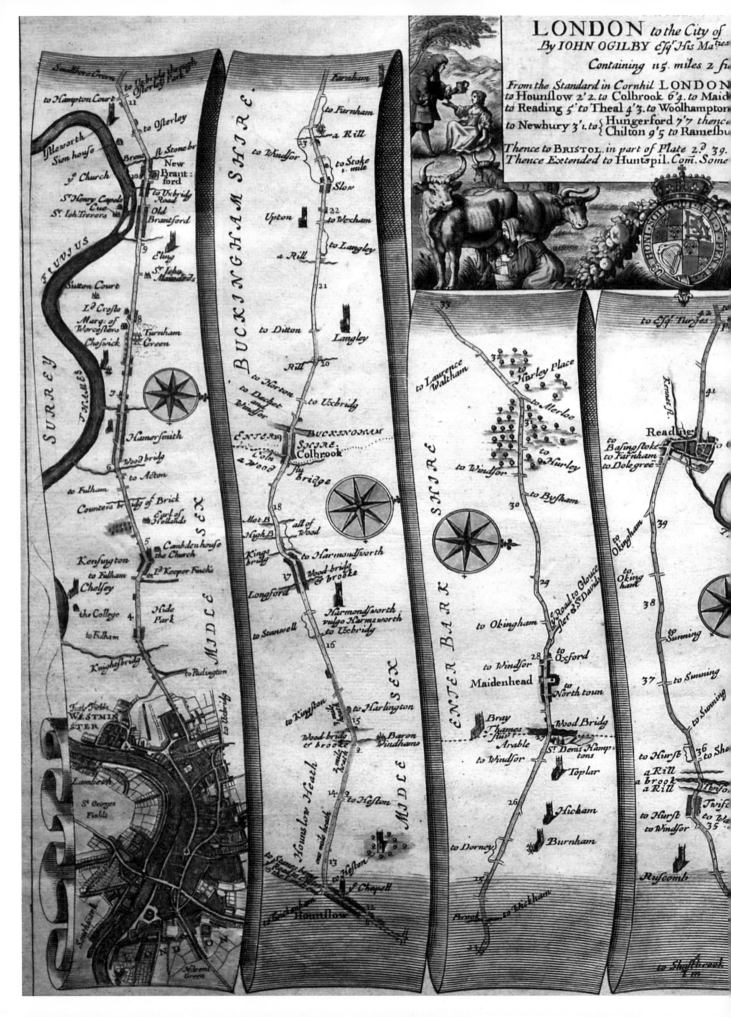

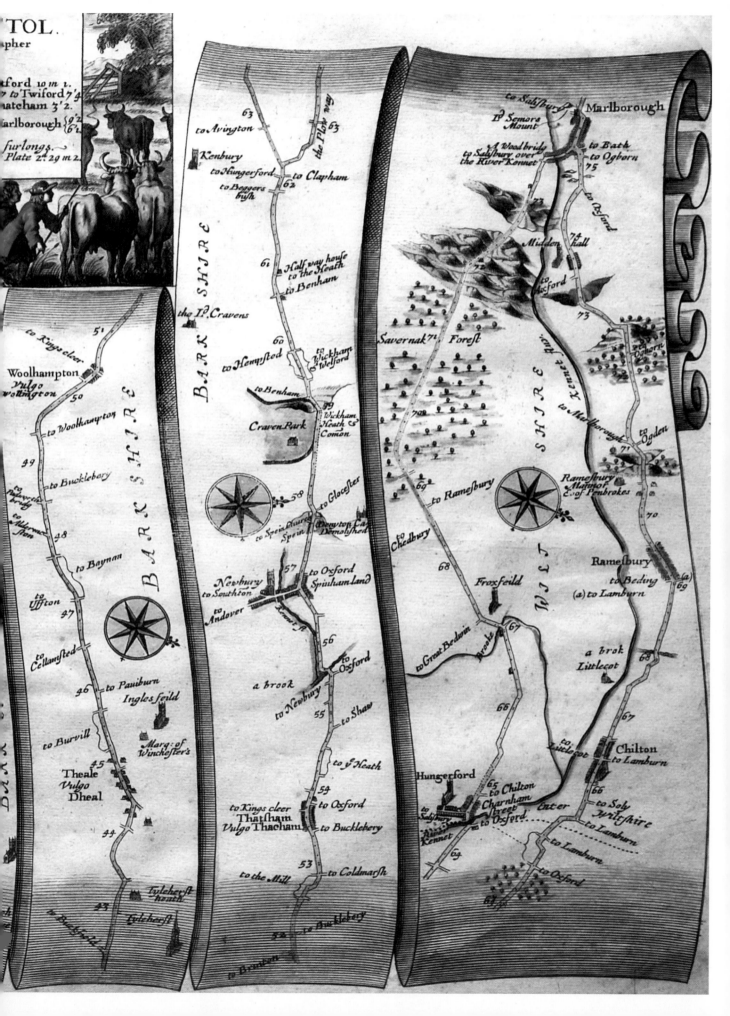

MAP OF HERTFORDSHIRE,
John Seller

As well as producing sea charts, John Seller started to map the counties of England. His 1676 map of Hertfordshire shows a dense network of towns and villages in one of England's more prosperous counties. Seller's work showed only incremental improvement on the work of pioneering county cartographers such as Christopher Saxton, an apt reflection of a climate in which the high hopes provoked by the Restoration had given way to disappointment at the political bickering, war, plague and fire which had afflicted the nation.

Hertfordshire could boast some of England's most ancient settlements. St Albans, or Verulamium, was a pre-Roman foundation, which was burnt by Queen Boudicca of the East Anglian Iceni tribe during her revolt against the Romans in AD 60–61. It was also the site of England's first recorded martyrdom, when the eponymous Alban was executed in 303 for refusing to perform a sacrifice to the Roman Emperor Diocletian. Further south, in 1471, Barnet had seen a decisive battle during the Wars of the Roses, when the death of the Earl of Warwick, who had recently defected to the Lancastrians, signalled the final victory of the Yorkist cause.

In 1661 Barnet also experienced a curious revolt, when fifty supporters of the Fifth Monarchy Men, a radical religious sect, attempted an uprising, which aimed to reach the centre of London and unseat Charles II. The Fifth Monarchists made little headway, were easily defeated and their leader Thomas Venner and nine of his supporters were hanged. They belonged to one of the splinter groups which had flourished in the fervent atmosphere of the English Republic, when the more extreme Puritans believed that the second coming of Christ was imminent and that a community of saints needed to be established on Earth.

Sects such as the Levellers, who believed in equality of all before the law, the Diggers, who espoused the common ownership of property, and the Fifth Monarchy Men, who considered that five monarchs would achieve earthly dominion before the ending of time (and that Charles II might well be the fifth of them), posed a considerable problem to the restored monarchy. By the Declaration of Breda, just before his arrival back on England

in 1660, Charles had promised clemency to most of those who had opposed him (including a fair clutch of religious radicals), but the Act of Indemnity and Oblivion which put this promise into law excluded fifty people, most of them regicides, who had signed the death warrant of Charles I in 1649. Among those whom the government could not arrest, for he was already dead, was Oliver Cromwell. In a fit of vindictive rage, Charles ordered that the Lord Protector's corpse be exhumed, paraded in public and decapitated.

The strongly Royalist Cavalier parliament of 1661 passed a series of measures aimed at dissenters, non-Anglican Protestants. The Corporation Act required those holding public office to swear allegiance to the crown (something many dissenters would not do). The Conventicle Act of 1664 went even further, by forbidding religious gatherings of more than five people. Ultimately, Charles wanted to relax existing restrictions on Catholics, while retaining them on the more radical Protestants, a policy which arose from his desire for a French alliance (cemented by the Treaty of Dover in 1670) and his generally pro-Catholic sympathies (his mother had been a Catholic and he had spent most of his long exile in Catholic countries). The governance of the country became bogged down in a tussle between Charles and parliament over the matter, which ended in the king backing down and acceding to the passing of the Test Act in 1673 which, by making it in effect compulsory to be an Anglican in order to hold public office, was almost precisely the opposite result to that which he had desired.

Although Catholicism was not fully tolerated until the Catholic Relief Act in 1829, most nonconformist Protestants were granted freedom of worship in 1689 when the Toleration Act removed restrictions on them as long as they made an oath of allegiance. For the Fifth Monarchy Men, however, that quixotic revolt in Barnet had been their last hurrah. Most of their leadership had been executed and when the magical year of 1666, which many of them had believed would mark the end of the world, passed uneventfully, the sect simply withered away.

HERTFORD SHIRE

Actually Survey'd and
DELINEATED
By John Seller
Hydrographer to the King
Cum Privilegio Regis

THE
EXPLANATION
OF THIS MAPP

ANGLIAE TOTIUS TABULA,

John Adams

A thick network of lies radiate in all directions on in this unique map by John Adams, a Shropshire lawyer turned mapmaker. They show the distance in miles between England's towns and villages, accompanied by a table (not shown here) giving the distance from London and latitude and longitude co-ordinates for each of them. Yet the dedication of this 1693 reissue to King William III ('Gulielmo III') was a sign of the great political changes that England was undergoing at the time; for the 1685 original version of his *Angliae Totius Tabula* ('A table of all England') was dedicated to Charles II, whose successor, James II, had been deposed by William in the Glorious Revolution.

If Adams needed to be cautious with his dedications, there was no doubt over the care he lavished on his great mapping project. He began it in 1677, but criticism over the number of settlements he left out of the first version led to a further survey which culminated in the publication of the *Angliae Totius Tabula* in 1685. It had all begun as a relatively simple idea; a friend who had fishery business in Aberdovey on the west coast of Wales had wanted to work out the market for his fish within a hundred miles and had asked Adams to help. The result was a table of distances to Aberdovey and the extension of this to London and beyond.

Another simple idea was growing in England and one whose antecedents were a century deep; that no Catholic should ascend the throne of England. By the mid-seventeenth century there were actually relatively few Catholics left in England; perhaps 70,000 or 1 per cent of the population still professed the old faith and they offered no real threat to the Anglican establishment. Yet Charles II's marriage to the Catholic Catherine of Braganza and his dabbling with a French alliance roused suspicions that he wanted to restore England's allegiance to the Papacy. Worse still, the rumours that his brother James, Duke of York, was a crypto-Catholic were confirmed in 1673 when he refused to swear an oath under the Test Act, a measure designed to prevent Catholics holding office or seats in parliament.

The febrile atmosphere of anti-Catholicism which now permeated the court and London society more generally led to the revelation by Titus Oates in 1678 of a widespread plot to kill the king and place James on the throne. This Popish Plot was in fact a complete fabrication, but it fuelled suspicion of James and, by extension Catholics, and led to an attempt between 1679 and 1681 to pass an Exclusion Act barring James from the throne.

Charles resolutely refused to disinherit his brother, and was, as a result, challenged by the Rye House plot in 1683 in which leading anti-Catholics, including the Duke of Monmouth (Charles II's illegitimate son) were caught plotting to kill him and place a more faithful Protestant (quite possibly Monmouth himself) on the throne.

When Charles finally died in February 1685, a revolt broke out in the West country led by Monmouth who had escaped to the Continent the previous year returned, landing at Lyme Regis in Dorset. Although he gathered several thousand supporters, the collapse of a parallel revolt in Scotland put him on the defensive and his army was cut to pieces at Sedgemoor on 6 July. The victory invigorated James to accelerate his plans for the restoration of Catholicism. In April 1687 he ordered clerics throughout the land to read a Declaration of Indulgence from their pulpits, which announced that henceforth toleration would be granted to all non-Anglican Christians (which included Catholics).

James's intemperate actions caused even moderate opinion to desert him, and the birth of a son to his Catholic queen, Mary of Modena, in June 1688 (thus ensuring a Catholic succession) proved the final straw. Seven leading politicians, including Edward Russell and the Earls of Devonshire and Danby wrote to William, the Prince of Orange, inviting him to take the English crown (to which he had reasonable claim through his marriage to Charles II's daughter Mary).

On 1 November, William set sail from Holland, his little armada carried westward by a 'Protestant Wind' that hampered the attempts of the English fleet to catch him and he landed four days later at Brixham in Devon. James, seeing his allies deserting him, panicked and turned his army back from its march to the West Country to intercept the invaders. Seven weeks later he was a fugitive in France, his throne occupied by the safely Protestant William.

Adams had never been incautious enough to dedicate an edition of his map to James II, but he was wise enough to do so to William. Mapmakers might devote themselves to charting the changes in the landscape of their country, but they need to be wise, too, to the vicissitudes of politics.

MAP OF POST ROADS OF ENGLAND, George Willdey

London sits at the centre of a spider's web of roads in this 1713 map by George Willdey. It demonstrates the preoccupation of the age with the delivery of the post, a pressing problem given the comparatively poor state of the country's roads and the growing demand for a letter service that would transport mail safely between its towns.

Governments in Britain had long enjoyed access to some form of official postal service, ever since the Romans introduced the *cursus publicus*, the network of messengers and station-houses that ensured that imperial edicts and officially sanctioned travellers could move around the empire swiftly and efficiently. The oldest surviving letters in Britain were found on wafer-thin tablets of bark unearthed at Vindolanda, a fort just south of Hadrian's Wall, and date from around AD 70; amongst the correspondents was Flavius Cerialis, the commander of the Eighth Cohort of Batavians, and another soldier whose friends back home sent him a warm pair of socks to help him survive the northern winter.

Medieval monarchs had royal messengers who carried edicts and ordinances to county sheriffs and other officials and who possessed the right to commandeer horses locally when needed (in 1392 the justices of the King's Bench in York who got stuck in mud near Wansbeck were attacked by angry local villagers when they ordered them to hand over several horses to help pull them out). A formal service only got underway in Tudor times when Henry VIII appointed Sir Brian Tuke as Master of the Posts responsible for the care of the royal mail. By 1528, the first post office had been set up near Lombard Street to cope with the volume of correspondence.

The royal mail, however, was just that, reserved for the use of the crown and its servants. It was Charles I's need to raise money when parliament would grant him none that led to the opening up of the post to the public in 1635. Charges were by distance and paid by the recipient, at a cost of 2d for a letter going less than 130 km (80 miles), up to four times that amount for one carried to Scotland. After a period in which parliament farmed out the mail, allowing the 'farmer' to take all the receipts in exchange for a fixed annual payment, the service was taken back into central control in 1655 under John Thurloe, Cromwell's spymaster, who found it a convenient way to have access to the mail of anyone he deemed subversive.

At the Restoration, the first Postmaster General was appointed with a specific remit to run the whole new network of mail routes which was opening up. Rates were now formally set for postage abroad; a letter to Dublin cost 6d and to Venice 9d – while more frequent deliveries were begun to outlying areas; Penrith and Lincoln now enjoyed a weekly post and Truro was lucky enough to have deliveries twice a week.

An attempt by William Dockwra to undermine the government monopoly by establishing a private postal service in London in 1680 was initially wildly popular. He promised that it would be 'to the great Advantage of Inhabitants of all sorts' and at a cost of merely a penny to deliver anywhere in London, it soon undermined the official post. The Duke of York, who had been granted the revenues from the post, was displeased and launched a series of legal actions which, as heir to the throne, he was bound to win. By 1682 Dockwra was out of business and postal income to the government was once again rising steadily, from £108,000 in 1693 to £140,000 by 1700.

By the time Willdey produced his map, new local routes had been established, so that a letter could go directly from Exeter to Bristol, rather than having to be carried via London (resulting in the postage going down from 6d to 2d). A new Post Office Act of 1711 established a General Post Office for Britain (absorbing Scotland into the network for the first time). Although postage costs increased (so that shorter distances now cost 3d), a customer could now send a letter to New York for the bargain price of a shilling.

There was still a long way to go before Rowland Hill's introduction of a universal penny post within England in 1839 (and the introduction of the Penny Black, the first postage stamp in 1840). The roads were still poorly maintained and it was not until mail coaches were introduced in 1785 that mail travelled at all quickly between major cities. However, Willdey's map, with the squares indicating the major nodes of the network (the big cities) and the ovals the lesser towns, with double lines standing for the postal roads between the main settlements, is a sign that the public post was here to stay.

THE ROADS OF ENGLAND ACCORDING TO Mr OGILBY'S SURVEY

GROSVENOR ESTATE

This plan of the aristocratic Grosvenor estate in 1723 shows London on the brink of an expansion westwards from its traditional urban core in the City. New urban geographies were being laid out, as London escaped from its medieval shackles and Britain's political landscape underwent a similar transformation.

The Grosvenor estate had its origins in the marriage in 1677 of Sir Thomas Grosvenor and Mary Davies, heiress to a tract of unpromising swamp land in what is now Mayfair and Belgravia. The plan shows the bulk of this dowry tinted in pink, surrounded by green spaces and fields stretching from the future Buckingham Palace to the River Thames at Chelsea. It lay unused, and rather neglected until the 1720s, when the growing prosperity of the London middle classes led to an equal clamour for better housing than the teeming tenements of the capital's traditional residential quarters could provide.

London's population was not growing at this stage in its history – barely budging from around 670,000 between 1700 and 1750 – but there was a land shortage on which to build new suburbs and those aristocrats lucky enough to own estates on the fringes of the settled area soon saw an opportunity for profit. The Earls of Oxford, owners of the Harley (later Portland) Estate in Marylebone, Lord Burlington (whose estate was centred around Piccadilly) and the Grosvenors began to lay plans for grand squares with new roads radiating from them, all on land leased out to builders (who never acquired perpetual title to the plots, which the estates retained as freeholders).

Work began on Grosvenor Square in 1725, starting with large terraces, which were then joined later by a jumble of neoclassical mansions that led an unkind observer to comment that the square was 'little better than a collection of whims and frolics'. The architects employed were talented, among them Colen Campbell, who worked on Grosvenor Square and later John Nash, who laid out the magnificent Regency terraces by Regent's Park.

That the aristocrats were able to concentrate on the pursuit of money was in large part thanks to a gentler political climate, as the ravages and rivalries of the Civil War faded into folk memory. The Glorious Revolution of 1688, in which the Dutch prince William of Orange was invited to unseat the suspiciously pro-Catholic James II, was achieved with a minimum of violence. The transition from the last Stuart, Anne, who died, with no surviving children, in 1714 to her German Hanoverian cousin George was carried out with almost no fuss at all. Indeed, George was so confident that there would be no opposition that he delayed six weeks before arriving in his new realm.

The politics of the time, although vitriolic, lacked the edge of violence of previous generations. The germ of a party system was emerging, as Tories (originally supporters of James II and later tending to conservative Anglicanism) and Whigs (of a rather more dissenting and liberal disposition) struggled to form governments which exercised real authority rather than simply being proxies for the sovereign's will. By 1721, the office of prime minister had emerged, with Robert Walpole as its first incumbent.

Almost the only clouds on the horizon were the continued claims to the throne by the Stuart descendants of the deposed James II. His son James Edward ('the Old Pretender') had come back to the notice of the British political classes as a possible contender to the throne on Anne's death, if only he would convert from Catholicism. James refused to do so, but, with French support, carried out an abortive rebellion in Scotland, landing there on 22 December 1715. But by then his cause was already doomed, as his supporters, the Jacobites, had already suffered twin defeats at Preston in England and Sheriffmuir in Scotland, and Louis XIV, the chief French proponent of a Jacobite restoration had died on 1 September. By early February 1716, the Old Pretender had cancelled a planned coronation at Scone and had taken ship back to France, never to return.

The next major crisis in George's reign was suitably prompted by greed. The national debt was burgeoning, a growth exacerbated by a perennial royal habit of authorizing expenditure without the taxation revenue to back it up, and by 1720 had soared to £51,000,000. In an age of mercantile speculation, a company stepped in with a tempting solution. The South Sea Company, founded in 1711, made several offers to take on the national debt in exchange for a monopoly of trade to South America, and in 1720 finally reached an agreement that it would take over £30,000,000 of the debt. Shares in the company soared to astronomical levels, and aristocrats and the middle-class alike piled in to fuel the frenzy. When the bubble inevitably burst as it was realized the promised trade was a mirage, thousands were ruined. Almost the only silver lining was that one man who had got out early with a handsome profit was Sir Thomas Guy, who used it to fund a new hospital.

Other, wiser members of the nobility had directed their resources more widely: men such as Richard Grosvenor, Sir Thomas Grosvenor's grandson, whose planning of the new estate secured prosperity for his family and made an imprint on London's urban landscape that has endured.

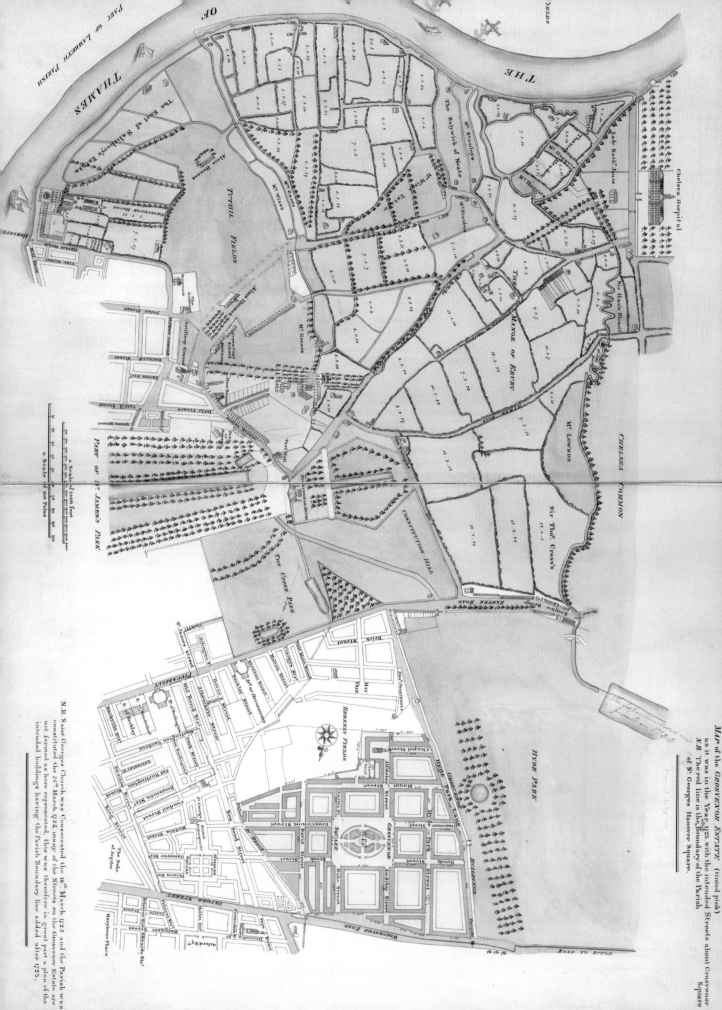

MAP OF SCOTLAND,

John Elphinstone

John Elphinstone's 1745 map of Scotland charts the country on the eve of the defeat of the Jacobite attempt to place the exiled Young Pretender, Charles Edward Stuart, on the British throne. The map's dedication to the Duke of Cumberland, the British government commander, whose forces used it during the campaign, gave it a retrospective stamp as a portent of doom.

Elphinstone was an engineer serving with Cumberland and his 'new and correct map of North Britain' was the most up-to-date available in the 1740s. Despite his evident pride in his work, however – his pointed annotation to the isolated Isle of Rona declares that it is 'omitted by all the map' yet 'every Navigator that has been in these Seas knows the Existence of them' – Elphinstone was subsequently criticized for the inadequacy of his depiction of the hilly topography of the Highlands, which did little to aid military route planning during the 1745–6 campaign.

The roots of the 1745 Jacobite uprising lay in Scotland's troubled relationship with the Stuarts while they were reigning and the determination of many to have them restored once they were deposed. Relations had got off to a rocky start after the Restoration of Charles II when the Act Rescissory of 1661 rescinded almost the entire legislative programme of Scotland's Covenanting parliaments. Many Presbyterian ministers were thrown out of their livings (274 between 1660 and 1688) and suspicion boiled over into outright revolt when James II introduced toleration for Catholics in 1687.

This rehearsal for the Jacobite rebellions of 1715 and 1745 was easily put down at Killiecrankie in 1689, but Scotland's discontent festered. A scheme launched in 1695 to plant a colony at Darien in Central America as a means of circumventing English restrictions on trade with the Americas (from which Scottish vessels were excluded) failed in a hell of heat and fever, bankrupting the investors (who had raised around £400,000) and delivering a near fatal blow to Scotland's economy. As a result, although the Scottish parliament elected in 1702 was initially hostile to England, resistance to the proposals for an Act of Union between the two countries crumbled when faced with the carrot of a bail-out for the Darien investors and the stick of the Alien Act, which threatened to bar the import of Scottish products into England if the union was not accepted.

Once the Act of Union was passed in 1707, London continued to neglect Scottish interests, pushing malcontents to renew plotting for a Jacobite restoration, a tendency reinforced when Queen Anne's death in 1714 opened up the very real prospect of a Stuart restoration. But the uprising of 1715 was ill-conceived and tentative. Its chief supporter, the Duke of Mar, hesitated to act decisively against the English and then achieved only an inconclusive result against the government forces under the Duke of Argyll at Sheriffmuir. By the time the Old Pretender, James Edward Stuart, arrived in December, the cause was already hopeless.

In the aftermath of the revolt, General Wade, the military governor despatched by the British government, built a network of military roads to make troop movements easier in the suppression of any future revolts. In the event they did little to stop the 'king over the water', the Old Pretender's son, Charles Edward Stuart ('Bonnie Prince Charlie'), gaining support when he landed in Scotland on 2 August 1745. He had come over in the wake of an Anglo-French war which had broken out in 1740. Initially intended to be used by the French as a pawn in a planned landing in southern England in 1744, he broke loose of his sponsors – whose promises of aid and men never seemed to materialize – and came in person to raise the clans.

At the start the response to his landing was indifferent, with the larger clans (save the Camerons) largely remaining neutral or actively opposing him, but his rapid capture of Perth, and then Edinburgh, brought waverers flocking to his standard. His army swollen with new recruits, the momentum of these successes and the able generalship of Lord George Murray brought a string of successes until the Scottish host found itself at Derby, just 210 km (130 miles) north of London. There the prospects of facing fresh British armies, reinforced by men returning from the war in Flanders, a possibly bloody fight over the south of England and sheer war-weariness, turned the tide in favour of caution and a rapid retreat northwards began.

The Jacobites were pursued all the way by the Duke of Cumberland and by early December had crossed back into Scotland. They then melted back into the Highlands, while Cumberland, pausing at Edinburgh to regroup, struck out in search of them in early Spring. At Culloden Moor, near Inverness, on 16 April 1746, the Jacobites' hopes were shattered; the Highland clans were cut to pieces by musket fire and the force of sheer numbers (double their own). Charles fled, a hunted fugitive until he took ship back to France five months later, and at his death in 1788, his brother Henry, a Catholic cardinal, became the Stuart heir, rendering the hope of a Jacobite restoration a sterile dream.

The Disarming Act of 1746 imposed penalties for the carrying of arms, while the wearing of tartans was banned and the jurisdictions of the clan chiefs were abolished. The final consequences of James I's journey down to London in 1603 had by now become clear, an eventuality that Elphinstone's description of Scotland as 'North Britain' made almost prophetically clear.

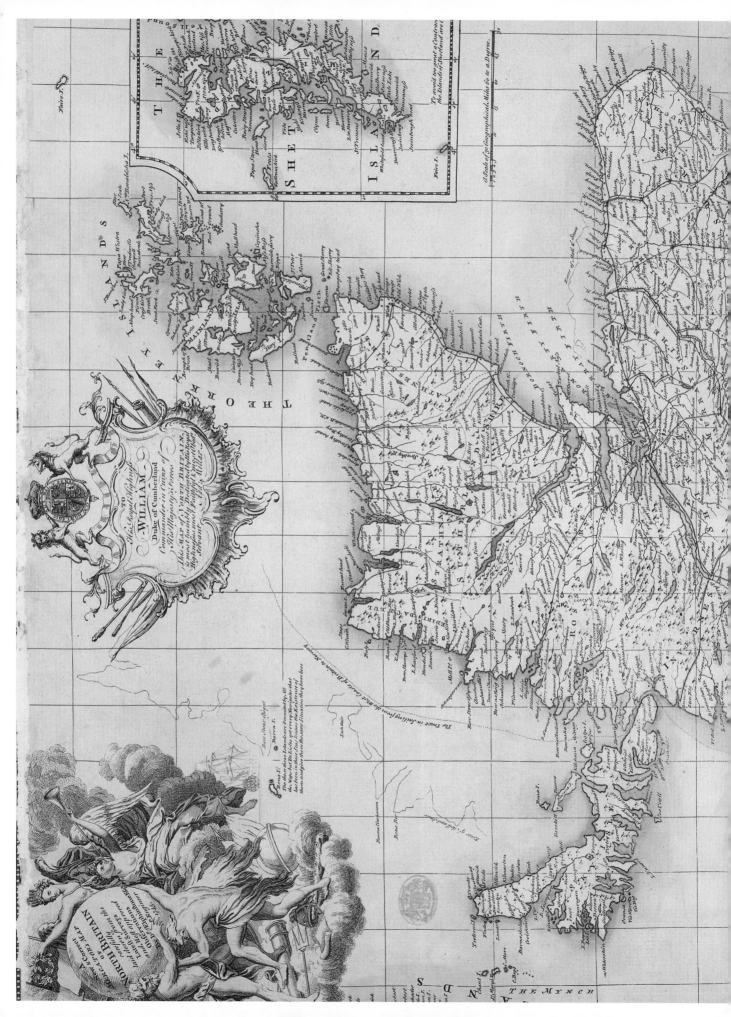

MAP OF LONDON, John Rocque

John Rocque's map of London, the most detailed yet produced, showed the city in 1746, in the middle of the reign of George II, a period of consolidation as new squares continued to be built on the edge of the metropolis and before renewed engagement in continental wars brought Britain the beginnings of an unparalleled global empire.

The map, with its detailed index entries and sure, confident engraving was surveyed by John Rocque, a French Huguenot emigré who had begun his mapping career producing estate plans for aristocrats, notably one of Richmond Gardens (now Kew Gardens). He started his survey of London in 1737, but so herculean was the task that it took him seven years to complete, and the result finally published in a 16-sheet edition.

He caught the city in a state of change; the aristocratic Harley, Grosvenor and Burlington estates having constructed a network of new streets and squares around Piccadilly and north of Oxford Street and High Holborn, areas now firmly part of London's inner city, but then lying at the edge of the surrounding countryside. The capital was a teeming, boisterous and occasionally dangerous place – moralists such as Hogarth depicted in searing prints the social iniquities caused by the craze for gin which exploded in the 1740s (to the extent that average consumption of the spirit reached two pints per head a week, or nearly 70 million pints a year).

There was social progress, too. New hospitals such as the Westminster General Infirmary were founded (in 1719) and the capital acquired Britain's first specialist maternity hospital, when the British Lying-In Hospital opened its doors in Covent Garden in 1749. Concerned over the decay of traditional morality, John Wesley, the founder of Methodism began preaching in London in 1738 (and purchased his own chapel, a disused Moorfields iron foundry, the following year).

Those with a taste for news had found it easier to satisfy their needs with the appearance of journals such as the *Daily Courant* in 1702, the *Evening Post* in 1706 and, eventually, *The Times* in 1785. Domestically, they would have read about a series of political crises as successive Tory and Whig administrations came and went (after the fall of the stabilizing hand of Robert Walpole in 1742). Less prominent might have been the quarrels in the royal family which had left George II barely on speaking terms with his father and resulted in a banishment from court in 1717 and a contremps with his own son, Frederick, Prince of Wales, that led to a decade-long exclusion from court.

The most dramatic event of the reign was the Jacobite uprising in 1745, when soldiers loyal to the 'Young Pretender', Charles Edward Stuart (the grandson of James II), reached as far south as Derby before retreating northwards and facing eventual annihilation at Culloden, near Inverness, in April 1746. Britain also became involved in a series of overseas wars, in part through the renewed engagement with European affairs brought through George I and George II's role as Elector of the German state of Hanover. First sucked into European monarchical rivalries in 1739 through the picturesquely name War of Jenkins' Ear against Spain (named for an unfortunate English sea-captain whose ear was sliced off by a Spanish patrol off the coast of Florida), the British fought in the subsequent War of Austrian Succession to defend the right of Maria Theresa to the Austrian throne. When news reached London in 1748 that the war had been won, the outbreak of peace was celebrated by a fair in Green Park for which Handel composed the 'Music for the Royal Fireworks'.

More serious in its long-term consequences for Britain (and for London) was the Seven Years' War which began when tensions between the British colonies in North America and the French holdings in Canada spilled over into open warfare. The conflict spread until it had become a global one, with theatres in the American East coast, the Caribbean, Europe and India. The British had a most successful war; Robert Clive's victory at Plassey in India in 1757 initiated the massive increase in the territories controlled by the British East India Company, while James Wolfe's defeat of the French general Louis-Joseph de Montcalm at the Plains of Abraham outside the city of Quebec two years later led to the British acquisition of almost all France's North American colonies.

These new territories would bring new trade to London, and the resulting prosperity would lead to a further expansion of its bounds. When John Rocque carried out his survey, though, only the prescient would have predicted how far those processes had yet to run.

see more on next page >

A PLAN OF THE CITIES OF LONDON AND WESTMINSTER,

WITH ALL THE NEW ROADS THAT HAVE BEEN MADE ON ACCOUNT OF WESTMINSTER BRIDGE, AND THE NEW BU

This PLAN extends from East to West near Six Miles, and from North to South a little more than Three, and contains about 11500 Acres of

ROUGH OF SOUTHWARK, AND THE CONTIGUOUS BUILDINGS;

AND ALTERATIONS TO THE PRESENT YEAR MDCCLV. ENGRAVED FROM AN ACTUAL SURVEY MADE BY JOHN ROCQUE.

ed by the Proprietors of the Original Survey JOHN PINE., JOHN TINNEY, and THOMAS BOWLES, according to Act of Parliament.

THAMES

MILE END OLD TOWN

MAP OF DERBYSHIRE,
Peter Burdett

Carefully, almost pedantically engraved, this 1767 map of Derbyshire won an award for its creator, Peter Burdett. Drawn to answer concerns that previous county maps were becoming inadequate, Burdett worked on the cusp of the Industrial Revolution, in a county where enormous changes in the textile industry were about to transform Britain's landscape for good.

Burdett's prize was awarded by the Royal Society of London, which had announced a scheme to encourage new county maps, given, as one eminent natural historian lamented, 'Our Maps of England and its counties are extremely defective'. He was very much a Derbyshire man, a prosperous member of a county set which included Earl Ferrers, and the painter Joseph Wright of Derby. Among the members of this group who sat for Wright was Richard Arkwright, a pioneer industrialist in the cotton industry, whose calm, assured and slightly self-satisfied face peers out of his 1784 portrait.

Arkwright, whose career had begun in rather unpromising fashion as a barber, had every reason to feel satisfied, for in 1769 he was awarded the patent for a new process that would revolutionize the textile industry. Already John Kay's flying shuttle (of 1733) had improved the efficiency of weaving by allowing the shuttle to be passed more quickly through the loom, and James Hargreaves' spinning jenny (in 1764) had multiplied the number of spindles which could be operated at one time. Arkwright took these advances and refined them, using water power to operate larger looms and his water frame which he designed to spin high quality yarn at optimum speed.

In 1771 he built a new factory at Cromford, near Matlock in Derbyshire, using water power for his water frames and looms and set about attracting workers by advertising that he would permit them a week off a year (an almost unprecedented perk). Cromford, the first large-scale factory, was a harbinger of the Industrial Revolution. It operated two thirteen-hour shifts a day,

with bells that rang at 5 a.m. and 5 p.m. to mark the changeover between them. So successful was Arkwright's system that before long he had built additional mills at nearby Masson and then a string of further factories in Lancashire, Staffordshire and Lanarkshire (in Scotland).

Arkwright's success attracted the jealousy of competitors who considered his patents were impeding progress by not allowing them to enjoy a share of the bonanza in the expanding textile industry. In the end a court struck down his patents in 1785, opening the way to a further flood of investment that sucked agricultural labourers into yet more factories. By then, new processes had opened the way to advances in other industries, as James Watt's steam engine (patented in 1775) permitted further automation by harnessing the power of steam (and ultimately giving birth to the transport revolution of railways powered by steam engines). Henry Cort's method for puddling iron, devised in 1783, made a purer form of the metal available to build the factories, railway track and machines for which Britain's growing industries thirsted.

The wide open agricultural pastures depicted on Burdett's map would shortly be enclosed, factories would dot the landscape and towns expand as labour flooded in from the countryside. Britain would become the world's pre-eminent industrial power, but at a price: the reverse side of Arkwright's apparent philanthropy in offering his workers paid holiday was a system where efficiency came at a terrible human cost. So much so that a series of Factory Acts had to be passed in the nineteenth century to moderate the excesses of factory owners: in 1833 one forbade the employment of children under nine years of age in factories and laid down that those between nine and thirteen should work a maximum of nine hours a day; while another in 1850 set the maximum working week for women at sixty hours. Progress had its price.

STOCKTON TO DARLINGTON CANAL MAP

The map shows one of the result of Britain's growing 'canal mania' in the late eighteenth century; a proposal for a waterway to link the northeastern towns of Stockton to Darlington (and beyond to Winston and the Auckland coal mines), as a means of transporting coal to feed the growing appetites of the factories that began to mushroom as the Industrial Revolution gathered pace.

Cotton, coal, steam and steel were the quartet which drove Britain's industrial transformation. The patenting of the waterframe by Richard Arkwright in 1769 enabled multiple spindles of yarn to be spun rapidly and evenly, tipping the balance in favour of factories and against the traditional artisan spinners. The technical advances in producing wrought iron by Abraham Darby II of Coalbrookdale in the 1750s and by Henry Cort in the 1780s meant high-grade iron (and soon steel) became plentifully available for industrial purposes for the first time. Powering many of the new factories which mushroomed were steam engines, their heritage dating back to the version patented by James Watt in 1781; and providing fuel for all this burgeoning enterprise was coal, principally from fields in the north-east.

Coal (and iron and cotton) were bulk commodities which were difficult, and expensive to transport by road. An obvious alternative was by water, but many of Britain's waterways were either not fully navigable along their entire routes or did not link with others, making it impractical to use water transportation except in particular circumstances and for short distances. As the Industrial Revolution's appetites for fuel and goods grew ever greater, however, small localized experiments in constructing new waterways were carried out.

The Irish parliament in 1729 established a Commission of Inland Navigation to promote such schemes, with a particular view to building a canal from the mining area of Tyrone to allow coal to be shipped to Dublin. They appointed Richard Castle as the engineer and, after five years of labour – though he took time out to serve a term as Liverpool's mayor – he had completed a 29 km (18 mile) stretch of canal with fourteen locks, which was arguably Britain's first modern canal. It was not emulated until 1755 when Henry Berry began work on the

18 km (12 mile) Sankey Brook Canal between St Helens and the Mersey, again intended as a conduit for coal. The Bridgewater Canal came next, built by the Third Duke of Bridgewater, who had seen the Canal du Midi in France during his Grand Tour in 1753, and wanted one of his own. Running from his coal mines at Worsley to Manchester, and with an extension to Runcorn, the canal was finally completed in 1776 under the guidance of its chief engineer, James Brindley, a former millwright.

Projects for canals now became all the rage and within eighty years of the Bridgewater Canal, the British Isles had around 8,000 km (5,000 miles) of canals. Much of the building was financed by private operators, such as the Stratford to Birmingham Canal, which raised its £120,000 capital mostly from small shareholders, including two coopers, three parsons, a plumber and a joiner. As canal owners were not allowed to run cargo boats themselves, the operating costs were defrayed by charging tolls for passage, an unsteady source of income which left many canal companies financially very vulnerable.

It was into this landscape that the proposal for the Stockton and Darlington Canal landed. It was designed by James Brindley, the engineer of the Bridgewater, with Robert Whitworth, and was commissioned by the coal merchants of Darlington who wanted their mines to be linked to the navigable Tees at Stockton. Yet when the report was published in 1770, it was not possible to raise the necessary finance; the area was too sparsely populated and there was simply not the demand to finance its construction.

Even though by 1790 the Mersey, Trent, Severn and Thames were all linked into a network of navigable waterways, the golden age of the canal did not last long, and it was the replacement of the unbuilt Stockton and Darlington Canal that killed it. For the latest transportation novelty, the railway, proved to be far more flexible than the canal barges – which had to be narrow because of the width of the waterways and so could carry a maximum of 30 tons load. Instead, railways filled the gap. The first public line opened in England used a route similar to that surveyed by James Brindley: in 1825 the Stockton & Darlington Railway opened and was soon satisfactorily hauling far larger loads of coal.

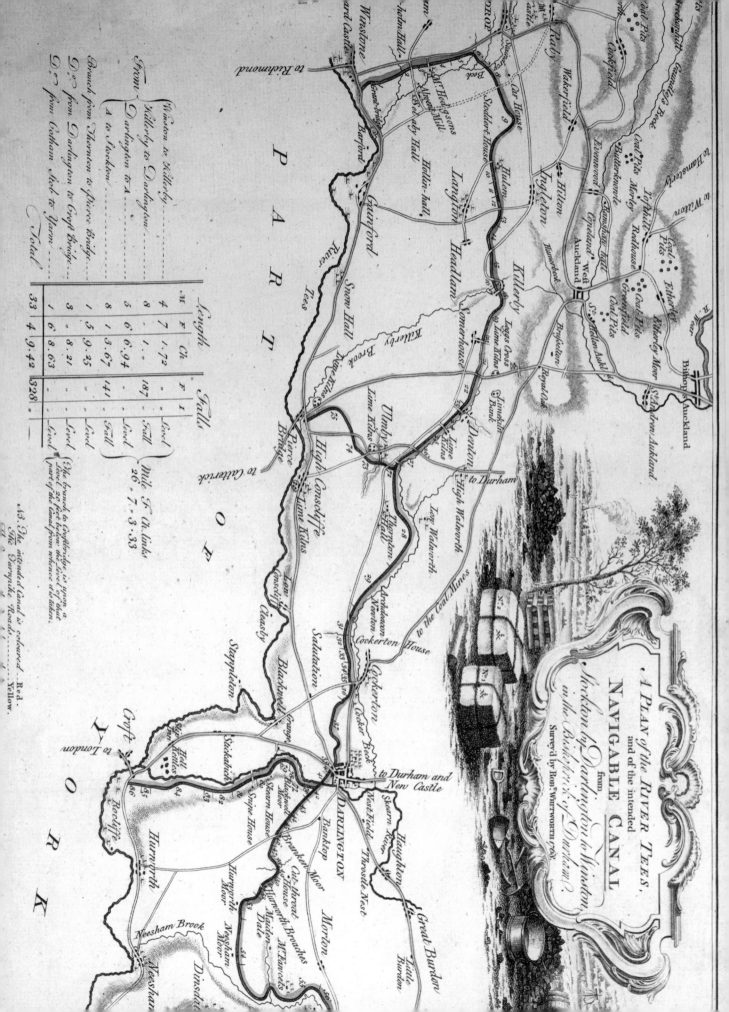

A PLAN of the RIVER TEES, and of the intended NAVIGABLE CANAL from Stockton by Darlington to Winston, in the Bishoprick of Durham. Survey'd by Robt. Whitworth 1769.

	Length			Falls	
	M.	F.	Ch.	F.	L.
Winston to Killerby	4	7	1.72	Level	
Killerby to Darlington	8	1	..	Fall	187
From Darlington to A	5	6	6.94	Level	
A to Stockton	8	1	3.67	Fall	141
Branch from Thornton to Pierce Bridge	1	5	9.35	Level	
Do. from Darlington to Croft Bridge	3	.	8.21	Level	
Do. from Cotham Fort to Yarm	1	6	8.63	Level	
Total	33	4	9.42	328	

Mile F. Ch. links
26.7.3.33

NB. The intended Canal is colowred — Red.
The Turnpike Roads — — — Yellow.

NB. The branch to Croftbridge is upon a
Level 20 feet below the level of that
part of the Canal from whence it is taken.

MAP OF LANCASHIRE,
Emanuel Bowen

The port of Liverpool sits prominently in the southwest of this 1777 edition of a map of Lancashire by the cartographer Emanuel Bowen. Yet already the prosperity of Lancashire's textile industry and Liverpool's trans-Atlantic carrying trade was being threatened by momentous events in North America, as Britain was about to lose a large portion of its infant empire.

Bowen had a long career as a mapmaker and the business passed to his son Thomas after his death in 1767. Although his most enduring creation was *Britannia Depicta*, a pocket edition of John Ogilby's compilation of route maps, the Lancashire map formed part of the *Large English Atlas*, a collection of county maps which was completed in 1760 and republished in several editions thereafter.

In 1777, Britain's army suffered a shattering setback in North America that was the harbinger of worse to come. The defeat of General Burgoyne (popularly known as 'Gentleman Johnny' for his sartorial flair) by American colonists at Saratoga marked a turning point in a war that would end with Britain's loss of thirteen of its American colonies (which then set about forming the nascent United States).

The conflict blighted the reign of George III, the third of the German Hanoverians to rule Britain, which had – at least as foreign conflict goes – begun rather successfully. When the Seven Year's War broke out in 1756, Britain became sucked into fighting on the side of Prussia, and then found itself ranged against France in what became the world's first truly global conflict, with theatres of war as far apart as Central Europe, the Caribbean, Canada and Bengal. The British enjoyed several glorious victories that became enshrined in the nation's martial mythology: that by General James Wolfe against the French at the Plains of Abraham just outside the city of Quebec in 1759 – during which both he and the French commander Montcalm died – allowed Britain to take over the whole of French Canada, while Robert Clive's defeat of the Nawab of Bengal at Plassey in 1757 permitted the British to embed themselves at Calcutta (now Kolkata) and began the gradual extension of the British Raj throughout India.

The war had been expensive, though, and if there was one thing the Tories and Whigs, who quarrelled their way through a series of unstable administration in the 1760s, could agree on, it was that the North American colonies should be made to pay their way. With a population that exceeded 1.5 million by 1750,

not only were the Thirteen Colonies increasingly prosperous, they were also becoming politically more assertive. When the British government imposed a series of taxes on the colonies, including the Sugar Act of 1764 and the Stamp Act of 1765 (which raised revenue by duties on publications such as newspapers), there were widespread protests. When a measure was passed in 1773 to help the ailing British East India Company by giving it a monopoly on the tea trade to North America and an exemption from import duties, the colonists reacted by dumping over 400 cases of tea into Boston Harbour.

A further British clampdown followed and with positions becoming irreconcilable, open conflict broke out at Concord and Lexington in April 1775. This 'shot heard around the world' was the prelude to a six-year war in which the British sought in vain to cut the colonies in two. The Americans suffered setbacks, but always fought back under the inspirational leadership of General George Washington. On 4 July 1776, the American colonies declared their independence, making the rift irreconcilable. Britain suffered from its relative isolation in Europe, with no firm allies, and when France intervened on the American's side, the French fleet was able to isolate the main British army in Yorktown, where on 19 October 1781 it surrendered. As General Cornwallis marched out with his troops, it is said that the military band played 'The World Turned Upside Down'.

George III's world was turned upside down, also, and his depression at the loss of his American subjects contributed to the descent into madness that overshadowed his reign after 1788. When lucid, he proclaimed: 'I shall never lay my head on my last pillow in peace and quiet as long as I remember the loss of my American colonies.' When less so, scurrilous gossip had it that he tried to shake hands with an oak tree, thinking it the King of Prussia.

Lancashire's world was upended, too, and that of Liverpool in particular. The city had become England's premier port for trade with the New World and a new wet dock had been built in 1771 for the importing of West Indian and American sugar and tobacco (and the export of cloth from the Lancashire mills). All this was cut off, and the situation was aggravated by the outbreak of the French Revolutionary Wars in 1792. It would be a long time before Bowen's Lancashire could right itself again.

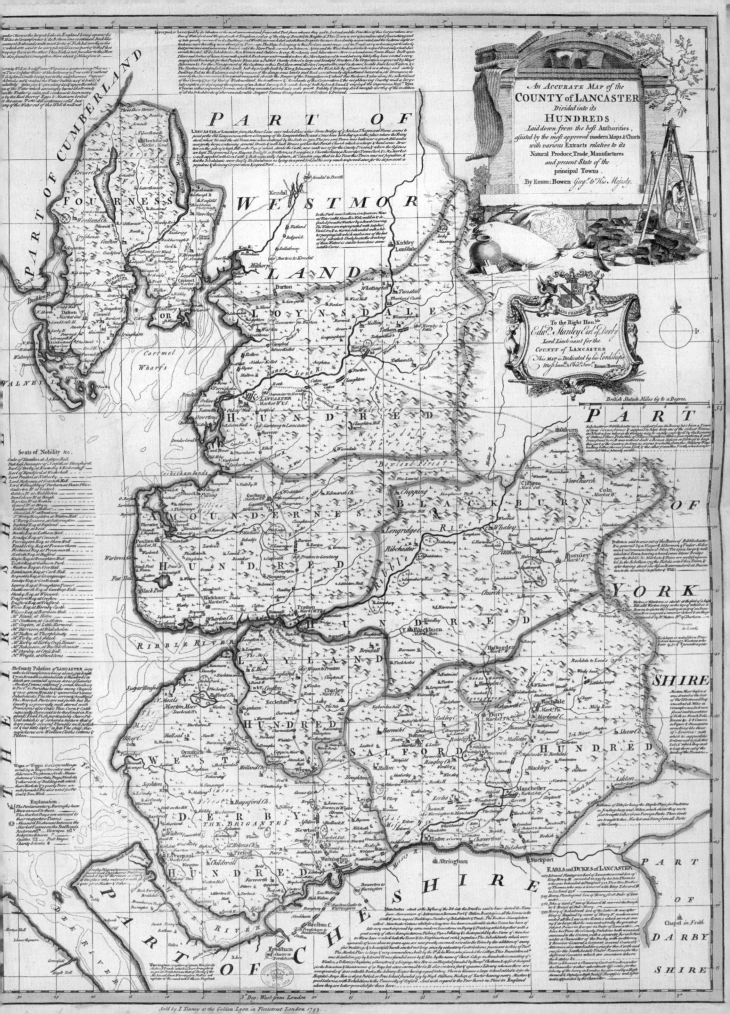

NORTHERN OCEAN

Orkney Isles
Pentland Firth
Wick
Ord Head
Dornock Firth
Murray Firth
Banff
Kinnairds Head
Buchan Ness
Aberdeen
Montrose
Firth of Tay
Firth of Forth
Dunbar
St Abbs Head
Berwick

Cape Wrath
Farr Bay
Johnny Groats
Loch Badcal
Loch Broom
Loch Ewe
Rona
Skye Isle
Grigadale
Mull Isle
Loch Levin
Kilmore
Jura Isle
Isla Isle
Jura Sound
Mull of Cantire
Firth of Clyde
Girvan
Fairland Point
Port Patrick
Mull of Galloway
Solway Firth

EDINBURGH

ENGLAND

CARICATURE MAP OF
SCOTLAND, Robert Dighton

This 1793 caricature map of Scotland, produced by the savage English social satirist Robert Dighton, shows a country 'bewitched'. The grotesque marionette-like figure that represents Scotland is red-faced and gaudily dressed, her tartan mantle creating an almost hunch-backed effect. It is a most unflattering, and perhaps unsurprising, portrayal at a time Scotland had suffered half a century of political neglect mixed with selective repression.

In truth, Dighton was not making a particular point about the Scots. He drew equally mocking caricatures of England and Ireland (the latter portrayed as Lady Hibernia Bull, the submissive wife of the 'great Mr John Bull'), in part as an aid to his publisher, Carington Bowles, who found them a useful way of encouraging sales of the more mainstream maps he peddled.

A chill had, however, descended over Scotland after the disastrous defeat of Bonnie Prince Charlie and his Jacobite army at Culloden Moor on 16 April 1746. Many of the defeated Jacobites were summarily shot and some of their leadership, such as the Earl of Kilmarnock, were beheaded. In its eagerness to ensure no repetition of the fright it has suffered when the Jacobites had reached Derby the previous year, the British government stationed twenty battalions of infantry and three regiments of dragoons in Scotland under the command of the Duke of Cumberland.

The Highland chiefs were stripped of their political authority and an act passed in 1746 banning the wearing of traditional Highland dress. The lands of those who had fought with the Young Pretender were confiscated and some 20,000 Highlanders emigrated between 1763 and 1773, many of them to the British colonies in North America. In place of the old chieftains, a cosy political oligarchy was set in place, whose compliant loyalty left little place for anti-English plotting. Chief among them was Henry Dundas, who became Solicitor-General at the tender age of 24, and who personally controlled twelve out of the forty-five Scottish seats in parliament, and whose influence over the rest was such, that he could muster a squadron of forty MPs to support the ministry of Lord North.

If politics stalled, learning did not. Scholarship, trade and the church (as well as service in the British army) were among the few options open to ambitious Scots, who could see little prospect of political advancement. The late eighteenth century saw a golden age for Scotland, which produced men of the quality of David Hume, whose *Treatise on Human Nature* became a classic statement of philosophical empiricism, the novelist Tobias Smollett, the diarist James Boswell and the philosopher and historian Thomas Carlyle. Perhaps the most influential of all was Adam Smith, whose *Wealth of Nations*, published in 1776 became a foundation stone of classical economics and attacked the prevailing mercantilist notion that an increase in trade (and so prosperity) could only be achieved by reducing that of other nations.

Scotland, though was beginning to be grow restive. In 1792, hardship caused by a bad harvest (and encouraged the publication in that year of Thomas Paine's revolutionary tract *The Rights of Man*) led to a demonstration in Edinburgh which burnt an effigy of Henry Dundas. Later that year a Convention of the Scottish Friends of the People was held in Edinburgh to lobby for electoral reform, though the English had lost none of their sensitivity about clamping down hard on trouble in Scotland and its chief organizer, Thomas Muir, was sentenced to fourteen years' transportation in the new penal colony of Botany Bay in Australia.

More seriously, the increase in the price of wool in the 1770s, when it nearly doubled, led landlords to privilege the grazing of sheep over the traditional rights of tenant farmers. Many had already migrated, leaving empty crofts, but many more were removed by force. These evictions were often brutal, pushing families onto marginal land from which it was hard to gain a living or rendering them completely landless. By 1811, some 15,000 tenants on the estates of the Duke of Sutherland alone had been evicted to make way for sheep. The memory of the Highland Clearances inflicted further bitter scars on parts of Scotland still smarting from the aftermath of the 1745 uprising.

After a visit by George IV in 1822, during which he wore a kilt, Highland dress became fashionable once more and experienced a complete rehabilitation during Queen Victoria's reign when a romanticized notion of Scotland's history took hold at court. In a way, the image which now prevailed was itself as much a caricature of the past as Dighton's misshapen creation.

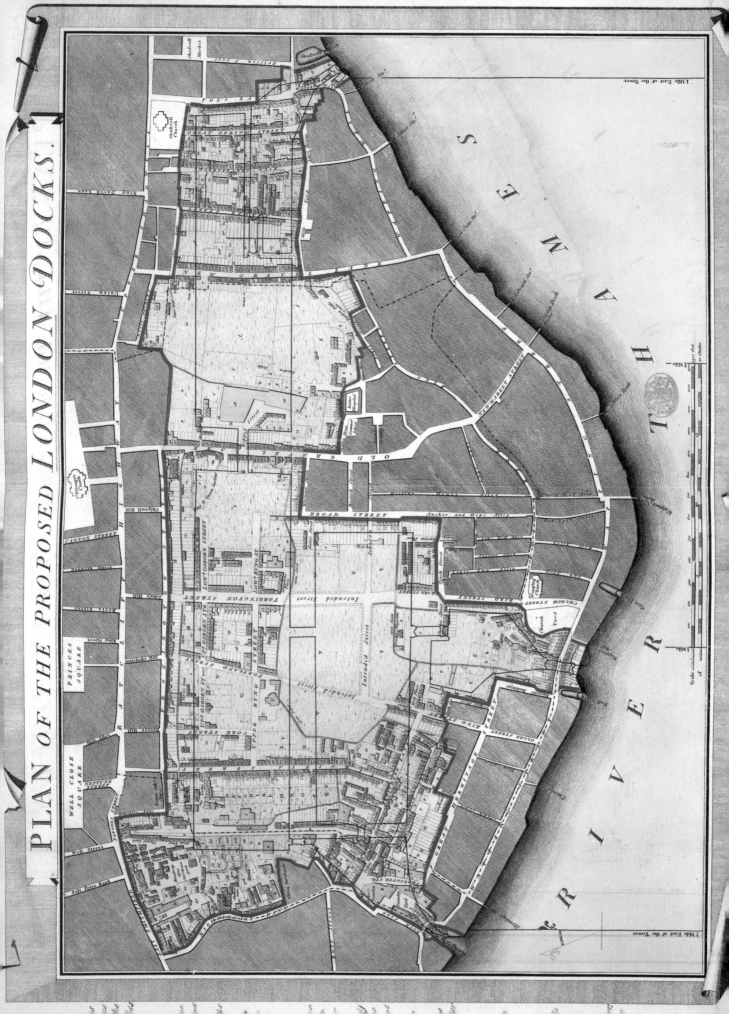

PLAN OF THE PROPOSED LONDON DOCKS.

PLAN FOR LONDON DOCKS

The pink-tinted area on this 1799 plan designated the proposed area for a new set of docks in East London, marking the start of a golden era in the city's career as an international trading port. Until their closure in 1969, London Docks would act as a prime conduit for the funnelling of goods into Britain from all over the world.

London had been at the heart of the nation's commerce since Roman times – as long ago as AD 61 Tacitus commented on the throng of merchants that visited it – but the topography of the river and the man-made choke point of London Bridge, which since 1209 had impeded the navigation of large vessels upriver, bequeathed it an enormous problem. No matter that in 1327 Edward III may have permitted foreign merchants to enter England in their own ships, or that the East India Company's foundation in 1600 brought a new surge of trade in luxury spices such as cloves and nutmeg, there was simply nowhere for all these ships with all those goods to berth.

As a result, apart from the wet dock built at Rotherhithe in 1696, the unloading of wares at London depended on a vast fleet of lighters and other small vessels. Queues built up, shortages arose and the backlog of goods mouldering on ships and piled high on wharves was subject to pilfering on an almost industrial scale by a small army of specialized thieves. Each year it was estimated between £250,000 and £800,000-worth of property was purloined by 'night plunderers', 'light horsemen', 'mudlarks' and 'scuffle-hunters'.

With trade into London reaching 3,663 ships in 1794, carrying goods worth £12.1 million (some 65 per cent of the English total), something had to be done. London was even falling behind other ports; while Bristol had 430 m (1,400 ft) of quays, London creaked along with only a quarter that amount and Liverpool had had its first proper dock before 1717.

The outbreak of war with Revolutionary France after 1792 proved a catalyst, as the fighting itself disrupted trade and raised the spectre of blockades. Powerful trading interests lobbied for the building of new docks for London, and in 1799 a Parliamentary Commission was established to hear evidence about a series of proposed new schemes. But powerful arguments for progress faced equally bitter opposition from an array of entrenched interests including City livery companies, Oxford colleges and the Company of Watermen which maintained that half its 12,000 members involved in unloading wares onto lighters could be out of a job.

Despite this the West India Dock Act was passed in 1799 giving consent for two new docks in the Isle of Dogs dedicated to trade from the Caribbean. A rival scheme, the London Docks, was passed the following year. The area selected at Wapping (depicted on the plan) would house up to 430 vessels, dedicated to high-value, luxury goods such as tobacco. The expense of construction was prodigious, inflated by the cost of the war with France and the need to buy out a large number of property-owners who owned houses on the site (as the explanatory text on the plan enumerates, some 7 ha (18 acres) of the site's 39 ha (96 acres) were covered by residential properties).

Finally, on 26 June 1802 the foundation stone was laid, buried with it a medal celebrating the Peace of Amiens, which it was hoped had brought an end to the war with France (and a concomitant upswing in trade), an expectation that proved illusory. Nonetheless the London Docks proved immensely profitable (paying out a 10 per cent dividend to its investors for many years), and was soon joined by an East India Dock (for the trade from India itself and the Spice Islands) and a complex of docks on the south bank of the Thames in Surrey.

When the first ship docked at London Docks in 1805, it marked the beginning of a century and a half when London truly was the world's pre-eminent port and, in the same year that Britain won a signal naval victory against a Franco-Spanish fleet at Trafalgar, of an era in which Britain's image of itself as a maritime superpower was, on the whole, matched by reality.

ORDNANCE SURVEY MAP OF KENT

With its intricate hill-shading and close attention to the delineation of roads and settlements, this 1801 map of Kent resembles more a Renaissance engraving than a humble county plan. Yet for all its merit as an artwork, this is a historic document also, representing the first sheet map published by the infant Ordnance Survey, at the start of a painstaking process which took seven decades to map the whole nation.

Like many maps, the Ordnance Survey's first product was born of war. This had seemed a distant prospect in the mid-1780s when the *Ancien Régime* in France was firmly ensconced and relations with the British were relatively good. In 1784, a controversy arose because of a suggestion in the mapping of the leading French cartographer Cassini de Thury that the Royal Observatory of Greenwich was misplaced on British maps. Overcoming the implied slight to national honour, the Royal Society's Joseph Banks commissioned the first triangulation survey of Britain, to measure a baseline and then survey a series of arcs, from which angles could be measured and therefore (using trigonometrical calculations) accurate distances measured.

The man chosen to carry out the measurement was William Roy, who had proved his worth on an ambitious project from 1747 to 1755 to create a more accurate map of Scotland following the Jacobite Rebellion of 1745. Beginning in July 1787 it took Roy just over a year to complete the succession of triangles down to the south coast and then to meet up with the French surveyors' triangulations. It proved the Observatory had indeed been wrongly sited in previous maps, but only by a little.

By now, though, there had been worrying developments across the Channel. On 14 July 1789, a mob had stormed the Bastille in Paris, a fortress used as a prison by the government. Decades of dissent over royal neglect of the provinces, high taxation, the effects of famine and the total detachment of the court from the country's problems boiled over into a revolutionary upsurge that soon swept power away from Louis XVI and into the hands of increasingly radical political movements. On 21 January 1793, a wave of shock and revulsion swept Europe's courts when it was learnt that Louis had been executed, and governments trembled lest the Revolution consumed them, too.

Already in summer 1792 the French Revolutionary army had declared war on Austria and Prussia (which counter-attacked only to be severely rebuffed). Britain feared a French invasion and the government needed to have better information about the landscape of the south coast of England; where troops might be quartered, the distance between towns and villages, the layout of roads, and on what beaches the French might land, if they did invade.

Charles Lennox, who had been appointed Master-General of the Board of Ordnance in 1782, decided that a new survey was urgently required. Roy had died in 1790 and so in July 1791 Lennox authorized the purchase of a theodolite, the most sophisticated surveying instrument available, at a cost of £373, and appointed Major Edward Williams and Lieutenant William Mudge of the Royal Artillery as directors of the new Ordnance Survey to create maps fit for Britain's defence.

The process was painfully slow; it took until 1801 before the first sheet of Kent – at a scale of a mile to an inch – was ready. It was frustrating for the government in London, which was all too aware that in October 1797 Napoleon, France's new military hero, had an army of 120,000 men poised on the north coast of France, ready to strike for England when the command came. Luckily it never did, as the troops were directed instead to a quixotic invasion of Egypt that ended in a disastrous capitulation to the British in 1801, while a French invasion force that was gathered in the Channel Ports in 1804 was similarly diverted to a campaign against Austria.

Britain engaged in successive coalitions against Napoleon, punctuated only by a year-long peace after the Treaty of Amiens in 1802. Always, the French general managed to evade defeat, disrupt the alliances against him or employ his military genius to destroy their armies. In 1811, the Board of Ordnance grew so alarmed that its maps might be offering military intelligence to the French that they were withdrawn from public sale and did not become available again until spring 1816, after the final defeat of Napoleon by the Duke of Wellington at Waterloo the preceding June.

It took a further fifty-four years for the Ordnance Survey to complete its mapping of Britain. By 1870, when it was finished, the country had weathered the Napoleonic Wars and the prospect of revolutionary contagion had been averted. The Kent map, though, stood as a reminder of a time when the nation was genuinely believed to be in mortal peril.

MARTELLO TOWERS MAP

A string of numbered dots stand guard along the south coast, in this 1867 map of the area around Romney Marsh on the Kent/ Sussex borders. They mark the sites of the Martello Towers, small coastal forts built at the height of the Napoleonic Wars to guard against a feared invasion from France which failed to materialize.

Small and round, like a set of game pieces plugged into gaps in Britain's defences, the Martello Towers owe their characteristic shape to a fort in northern Corsica, assaulted by the British navy at the start of the French Revolutionary Wars in 1794. It took a serious cannonade and two days of bitter fighting to take the position and the commander of the attack, Lieutenant-Colonel John Moore (who died heroically at Corunna fifteen years later), took note of how the design of fort had allowed a small number of defenders to resist so long. He also remembered its name: Fort Mortella.

Britain had been at war with France for just over a year by then, joining the other principal European powers, Austria, Russia and Prussia in an attempt to contain the revolutionary contagion which had first burst out in the storming of the Bastille prison in Paris in July 1789, and which, by the time Mortella was taken, had led to the guillotining of Louis XVI. It was an act which had profoundly shocked the rest of the continent's monarchies, and obliged Britain to join in the first of what were eventually seven coalitions against France (from 1799 led by Napoleon Bonaparte as First Consul, and then from 1804 as Emperor). Whatever the vicissitudes of the war on land – and Napoleon fought a series of strategically brilliant campaigns, consistently defeating and dividing his enemies – Britain always had the sea as its shield and the navy as its sword.

British complacency was shaken by a French naval fleet which reached Fishguard in Wales in February 1797, disembarking 1,400 troops of La Légion Noire (so-called for their captured British army uniforms, which were dyed dark brown or black). Unfortunately for their leader, the Irish-American Colonel William Tate, heavy winds had forced him to abandon plans to arrive in Bristol and so seize a major town. Instead he would have to fight his way from the far west of Wales and his scratch force, many of them freed convicts, took the earliest opportunity to engage in a riotous bout of looting and heroic drinking. Dispersed and helpless, they were easily rounded up by local militia and the last French invasion of Britain ended in ignominy.

No-one knew at the time it was the last invasion. When war broke out again in 1803, following a year-long pause in the fighting, the assembly of a huge army in northern France (which by August 1804 amounted to 167,000 soldiers) seemed to confirm fears that Napoleon was about to invade. A Militia Act had been passed to allow 50,000 men to be called up and hundreds of thousands more joined up to a volunteer force. Naval forces were sent to attack French outposts in the Caribbean and diplomatic envoys urgently dispatched to build a new anti-French coalition.

Minds turned, too, to the pressing need to upgrade the south coast's defences. In summer 1803, Captain William Ford, with the support of General Moore, put forward a proposal for a chain of coastal fortresses. To be built at the bargain cost of £3,000 each, it was only natural that they came to be called 'Martello' towers, in a slight corruption of their archetype which Moore had stormed ten years earlier.

Military and parliamentary bureaucracy moved painfully slowly and it was the spring of 1805 before work began on building the first towers. Around 40 feet high with a round shape that allowed a 360-degree gun traverse, they were each to be garrisoned by an officer and twenty-four men. The forts were to run along the south coast from Folkestone in Kent to Seaford in Sussex (with further towers eventually being built along the east coast as far north as Aldeburgh, in Suffolk, and in Wales and Ireland).

By the time the seventy-five south-coast towers were complete in 1812 (together with the Royal Military Canal built in 1804 to allow any invading force in Romney Marsh to be cut off), the threat of invasion was long over. Napoleon had been so confident that it would go ahead, that he had ordered the casting of a campaign medal celebrating 'La Descente en Angleterre' which showed Neptune (representing England) being crushed by a personification of France. Instead, further campaigns in central Europe consumed his attention, as did, from 1808, the growing ulcer of the Peninsular War campaign in Spain, where British troops slowly and inexorably pushed towards the French border.

The Martello Towers never fired a shot in anger. There were occasional proposals to revive them when subsequent waves of invasion paranoia swept the country and some were modified for use as coastguard stations to clamp down on smugglers. But for the most, they were left to moulder, crumbling concrete and brick testimony to an invasion which never came.

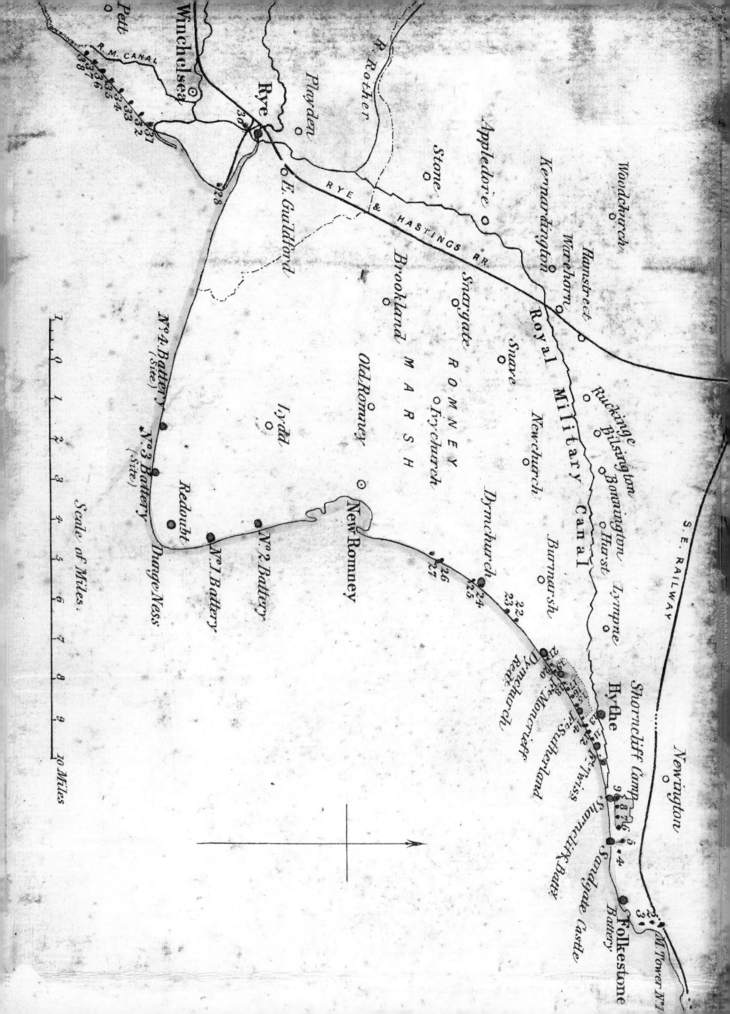

OF THE
STRATA
OF
ENGLAND AND WALES,
WITH PART OF
SCOTLAND;
EXHIBITING
THE COLLIERIES AND MINES,
THE MARSHES AND FEN LANDS ORIGINALLY OVERFLOWED BY THE SEA,
AND THE
VARIETIES OF SOIL
ACCORDING TO THE VARIATIONS IN THE SUBSTRATA,
ILLUSTRATED by the MOST DESCRIPTIVE NAMES
BY W. SMITH.

THE

GERMA

OCEAN

THE

IRISH SEA

FIRTH OF FORTH

FIRTH OF CLYDE

St. GEORGE'S CHANNEL

CAERNARVON
BAY

CARDIGAN
BAY

BRISTOL CHANNEL

EXPLANATION

ENGLISH CHANNEL

GEOLOGICAL MAP OF ENGLAND AND WALES,

William Smith

Rocks are carefully gradated and geological layers diligently shaded in William Smith's 1815 map of England and Wales ('with part of Scotland'). It was the culmination of a new approach that had begun with the dawn of the Enlightenment in mid-seventeenth-century Europe, a time which gave birth to a practical approach to the gathering and analysis of scientific data. No longer was the landscape thought of in terms of divine handiwork, but as a product of long, slow eons of sculpting by wind, rain and ice.

William Smith's comparatively humble background (born in 1769 to the village blacksmith at Churchill, in Oxfordshire) proved no obstacle to his establishing himself as the founder of modern British geology. Smith spent much of his early career working on the canals which were proving themselves as vital arteries along which coal and iron, essential for the Industrial Revolution, could travel. As chief engineer to the Somersetshire Coal Canal from 1794 to 1799, he surveyed a route from the coalfields to the markets in large towns such as Bristol. He noted the layers, or strata, in which rocks were formed and that in coal-bearing formations the same sequence of these repeated: sandstone, siltstone, mudstone, non-marine, marine and coal. He also observed the fossils found in any particular stratum were similar and by 1799 had devised a way to identify strata from the type of fossils found within them (which he finally published in 1816 as *Strata Identified by Organized Fossils*).

Smith began to travel throughout England and Wales, taking notes of the rock formations and layers he found on his journeys. He produced his first small-scale geological map, of the area around Bath, in 1799, hand-colouring portions to indicate the different rock types (yellow for oolite, blue for lias). Smith next produced a small-scale map of England and Wales in 1801. Yet this was not nearly the full extent of his ambitions

and for the next fourteen years he laboured on a far grander project, a map measuring 2 m by 3 m (6 ft by 9 ft) in 15 sheets, each of which took him over a week to hand-colour, showing in detail the geological composition of England, Wales and part of Scotland.

The results of decades of diligent work with pickaxe and notebook, Smith's work earned him the nickname 'Strata' Smith and a reputation as the Father of Geology, but he gained scant recognition in his lifetime. There was still a cleft in early nineteenth-century science between the scientific establishment and humbler, more practical exponents such as Smith. The Royal Society, founded in 1660, had provided an arena for the grandees of physics, mathematics and philosophy, men such as Isaac Newton and Robert Boyle, but its doors were not open to the autodidacts, engineers and inventors who observed and sifted through evidence gathered in the field. Similarly, the Geological Society of London, originally founded in 1807 as a dining club (and including amongst its founder members the chemist Humphry Davy), was not immediately receptive to Smith's ideas. Ignored and the victim of an unwise investment in a quarry near Bath, he ended up briefly in a debtor's prison in 1819 before resuming a diminished career as an itinerant surveyor.

Only in 1831. eight years before his death, did the Geological Society of London award Smith the first Wollaston Medal, its highest accolade. If Smith, the prophet of geological studies in Britain, had little honour in his homeland, he did at least receive extra-terrestrial recognition, when in 1987 a small crater in the Taurus-Littrow Valley on the Moon was named for him. As the site of some fascinating geology caused by the effect of a long-ago impact by a meteorite on the lunar surface, Smith, no doubt, would have been pleased.

see more on next page >

ST. GEORGE'S CHANNEL

CAERNARVON
BAY

CARDIGAN

BAY

BRISTOL CHANNEL

CREWKERNE ENCLOSURE MAP

The carefully drawn lines in this 1823 map of part of Crewkerne in Somerset mark an agricultural revolution. The award by commissioners of a section of local common land to large local landowners was part of a process by which enormous tracts of Britain were enclosed, virtually ending the centuries-old tradition of common access to agricultural land.

The medieval practice by which land was divided up into large numbers of separate strips (to which all had access after the harvest to graze their cattle and sheep) spread the risk of one field having a poor yield. In addition, the system of common land gave the poor some hope of rising above mere subsistence by growing their own vegetables or raising a pig. The system, though, was inefficient and did not allow larger landowners to consolidate their holdings, or permit them to implement agricultural improvements, such as allowing fields to lie fallow in order to regain fertility.

Some land had been enclosed – literally fenced off – in the Middle Ages. The process gathered pace slightly in Tudor times, particularly as demand for English wool made sheep farming more profitable than arable and led to the rich and powerful encroaching on common land or forcibly consolidating holdings to provide grazing for larger flocks. There was sporadic resistance, as existing peasant tenants were either forced off their land completely or provided inadequate exchanges with marginal areas, such as marshland. Kett's revolt in Norfolk in 1549 and the Midland revolt of 1607, led by 'Captain Pouch', an itinerant tinker from Leicestershire, were signs of deep rural anxiety about enclosure.

The pace of change, though, was very slow. The process to ensure legal title to enclosed land was cumbersome and time consuming. To resolve this, a large number of Acts were passed in parliament, each dealing with areas as small as part of a parish, which appointed commissioners who would make a decision on the reallocation of fields in the area and whose awards had the force of law. Between the 1730s and 1754 only four parliamentary enclosure acts were passed each year, but then, as low interest rates and higher food prices made land a more attractive investment proposition, the rate accelerated dramatically until, between 1790 and 1819, it reached seventy-five a year (helped by the upwards spike in food prices during the Napoleonic Wars, when importing food became more difficult in the face of French naval blockades).

In total around 2.8 million ha (7 million acres) were enclosed, representing about 23 per cent of the land surface of England. The majority of this land was in lowland areas of the south and southwest of England, where the old open field system had been most prevalent, but elsewhere common land, too, was enclosed, sometimes covering heathland and marshland. By the time General Enclosure Acts were passed in 1836, 1840 and 1845 mopping up remaining areas of common land, there was little left to enclose and the final act was actually intended to protect what remained in the form of public parks.

Large landowners were the big winners from enclosure. They could afford the additional costs of providing fencing around the new larger fields, while smaller farmers sometimes had to sell up a portion of their land to finance this (even though whatever they succeeded in keeping rose significantly in value). Many poorer tenant farmers lost out completely, as they were shifted onto far less productive land. Although there was no general exodus of labour from country areas, the rise in population (from 697,000 families employed in farming in 1811 to 761,000 families employed twenty years later) meant there was increasing pressure on resources, and the safety valve provided by common land had been removed.

The Crewkerne map, showing the enclosure of Roundham Common, illustrates these trends. The main portion of the previously common land is allocated by the Enclosure commissioner to two large landowners: the Hussey family, whose holdings around Crewkerne yielded above £3,000 a year; and John, Fourth Earl Poulett, a descendent of one of Charles I's leading supporters in Somerset during the Civil War. The poor, who watched as much of the benefit of the Agricultural Revolution accrued to their social superiors, must have reflected on the sentiment expressed by the Northamptonshire peasant poet John Clare 'Inclosure, thou art a curse upon the land, And tasteless was the wretch who thy existence plann'd.'

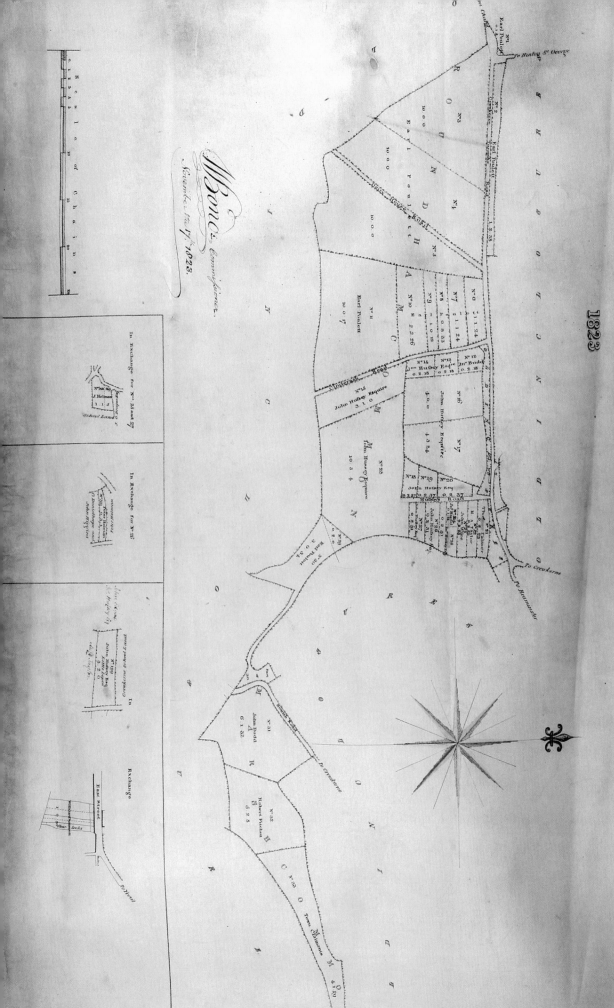

CREWKERNE INCLOSURE

AWARD MAP

1823

THE GREAT REFORM ACT

The streets and wharves of South Queensferry, northwest of Edinburgh, are carefully laid out in this 1832 map. Only the red line encompassing the town centre is a clue to its being the sign of a political revolution. It shows the new boundary of the constituency determined by the commissioners sent to survey Britain's cities and towns in preparation for the passing of the Great Reform Act which greatly expanded the electorate able to vote in parliamentary elections and swept away a large number of constituencies which had hardly any electors at all.

By the early nineteenth century, Britain's electoral system resembled the medieval heart of many of its cities; unplanned, chaotic, full of short-cuts, blind alleys and with plenty for sale to those with money. After the Act of Union with Ireland in 1800, the House of Commons had 658 MPs (with 512 for England and Wales). Representation was divided between counties (which generally returned two MPs, save Yorkshire which had four). Boroughs generally elected two members, although many, such as Abingdon and Monmouth, had to make do with one, while the City of London had four (and, unfathomably, little Weymouth had the same). Many growing cities, such as Manchester, were grossly under-represented while, others, such as Winchelsea and Dunwich, which had been important in medieval times, still had two members, despite their decayed state (and Dunwich had by 1800 largely fallen into the sea).

Corruption was rife. The enfranchisement qualifications were complex and confusing, in some boroughs related to property values, in others to income, in yet others being open to all 'freemen' (a status that could be bought at a price; in 1830 the corporation of Leicester hiked the price from £30 to £50 to raise extra cash in an election year). In 'rotten boroughs' such as Old Sarum, there were only 13 electors, many non-resident, who profited mightily at election time by selling their votes. In other 'pocket boroughs' there were in effect no electors at all, as a single patron controlled all the votes.

Calls for reform had got nowhere for decades and the Napoleonic Wars exercised a chilling effect on notions of change, as any whiff of radicalism was easily confused with (or accused of) revolutionary zeal. By the 1820s, though, momentum was growing. The Catholic Emancipation Act of 1829, which removed most legal restrictions on Roman Catholics (including allowing them to vote), split the Tory party, some of whom had supported the measure and gave the more reform-minded Whigs and Liberals an opening. In 1830 a new government took office under Earl Grey and, spearheaded by Lord John Russell began to push through measures to expand the franchise. Two bills failed in 1831, but a new election, fought mainly on the issue of voting reform, brought a larger Whig majority and the final passing of the Great Reform Bill in March 1832.

Fifty-six constituencies were abolished and thirty lost a member. These were redistributed to the more grossly under-represented towns and cities. The franchise was radically simplified, becoming open to anyone owning a property of £10 or more or paying a rent of £50 or more. At a stroke the electorate was increased by some 50 per cent, from 400,000 to about 650,000. Although this only represented one in five of the adult male population (and no women at all), it was at least a first step. In Scotland, for which a separate act had to be passed, the situation had if anything been worse. In Edinburgh, a city of 162,000 people, the corporation of just thirty-three members had elected an MP, while the country's entire electorate had amounted to just 4,329 men. The Reform Act increased this to 65,000, which was only around one in eight of the country's adult males. As in England, the property qualification meant commissioners had to be sent out, to carry out surveys of each constituency's land and buildings, with a value attached to each, to determine the owner's eligibility to vote.

There was still much to do. Further Reform Acts in 1867 and 1884 extended the franchise yet further, but universal male suffrage only came in 1918, at the same time married women over 30 were given the vote. By the time all adult women were enfranchised in 1928, nearly a century had passed since the Great Reform Act. The red band around Queensferry, however, suggests that a line had been crossed in 1832 and after that, the sweeping away of Britain's archaic electoral system so that ultimately all adults could vote, was only a matter of time.

QUEENSFERRY. (SOUTH)

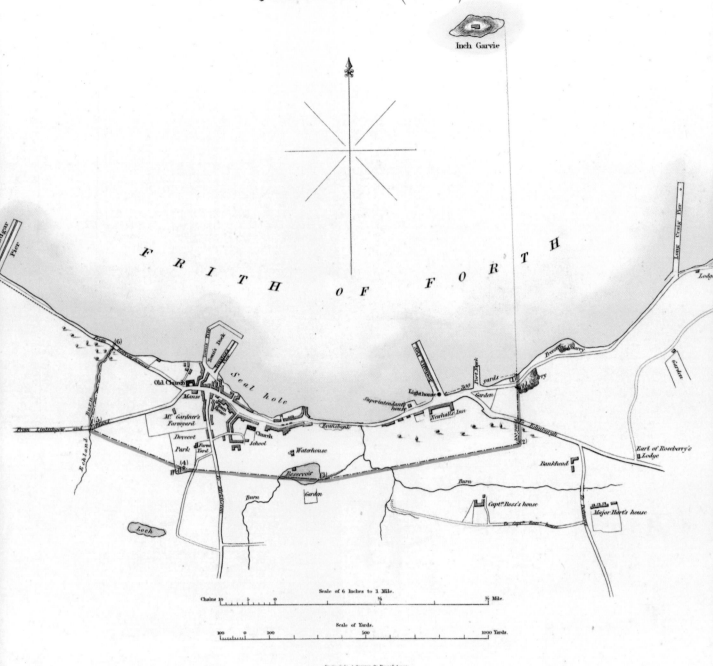

Inch Garvie

FRITH OF FORTH

Scale of 6 Inches to 1 Mile.

Chains 10   5   0   ½   ½ Mile.

Scale of Yards.

100   0   100   500   1000 Yards.

Davies sculp. 4 Compton St. Brunsk. Square.

LEICESTERSHIRE

TAMWORTH

Fazeley

ATHERSTONE
109

ROMAN   WATLING   STREET

Wibtoft

Swinfor

LUTTERWORT
90

Middleton

NUNEATON
105

Monks
Kirby

Caltthorpe

Church
Over

Chilvers
Coton

Bulkington

Withybrook

Harborough
Magna

Ansley

Bedworth

Shilton

Newbold

Clift
80

Over
Whitacre

Arley

Ansty Brinklow

Combe
Abbey

RUGBY

SUTTON
COLDFIELD

Shustoke

Fillongley

Corley

Sow

Church
Lawford

Bilton
85

Popes
Hayes

Water
Orton

Coleshill
104

Maxtoke

Allesley

COVENTRY
91

Binley

Brandon

Wolstone

DUNCHURCH
Wills

Erdington

Castle
Bromwich

Little
Packington

Ryton
on
Dunsmore

Streeton
on Dunsmore

Frankton

Leomillgton

Hastings

BIRMINGHAM
110

Yardley

Stone
Bridge

Meriden

Dunsmore

Bagington

Bubbenhall

Marton

Wapenbury

Aston

Sheldon
105

Bickenhall
100

95

Stockton

Edgbaston

Elmdon

Barston

KENILWORTH
101

Stoneleigh
Abbey

Weston

Long
Itchington

83
SOUTHAM

Moseley

Hall
Green

Knowle

Solihull

Temple
Balsall

Honily

Cubbington

Otchurch

Milverton

Radford
Simile

Ufton

H
Ladb

Kings
Norton

Baddesley
Clinton

Haseley

Hatten

Leamington
Priors

Whitnash

Harbury

Northfield

Lapworth

WARWICK
93

R

Preston
Pagot

Beaudesert

HENLEY
IN ARDEN
101

Oldberrow

Wootton
Wawen

Inslev

Redditch

WORCESTERSHIRE

TRING SUMMIT

| 1 in 330 | Level | 1 in 330 | 1 in 523 |

75   80   85   90

| Level | 1 in 660 | 1 in 1582 | 1 in 330 | Level | 1 in 330 | COVENTRY Level | 1 in 660 |

5   10   15   20   25

RAILWAY MAPS, James Drake

James Drake's 1838 maps of the London & Birmingham Railway and the Grand Junction Railway show Britain at the start of the railway frenzy, a rapid spread of the new means of transportation which enticed investors to fund lines in a spreading network of competing providers which soon touched almost every corner of the realm. The section shown opposite is part of the London & Birmingham Railway between Birmingham and Rugby. The Grand Junction Railway then permitted journeys north from Birmingham to Warrington, where connections could be made to Liverpool and Manchester. These routes were among the first of the true inter-city lines in Britain. The complete maps are shown overleaf.

Drake was a Birmingham engineer, whose maps provided an enormous service in helping passengers (and decision makers in parliament, who needed to approve new lines) make sense of the chaos of competing companies and routes which had sprung up. The London & Birmingham, stretching from the capital to Birmingham's Curzon Street station was fully opened in September 1838, while the Grand Junction, to which passengers could transfer in Birmingham, began full operations a year earlier. In these maps, Drake also portrayed the Euston Arch, the grand neo-classical monument built as a gateway to Euston Station (where it stood until its demolition in 1961–62) and depicted a post carriage (railway operators were from 1838 required by law to guarantee the passage of mail by railway) and carriages he imagines divided on political grounds between 'Reformer' and 'Conservative'.

Public railways were already more than thirty years old. Originally built to serve coal and iron or mining areas, the first to be built in England was the Surrey Iron Railway from Wandsworth to Croydon which opened in July 1803 to service oil-cake mills and move iron from barges on the Thames. By 1807 a prototype passenger service had begun on the Oystermouth Railway in South Wales. Yet these were only tentative steps and the first major passenger service was the Stockton & Darlington Railway which commenced operations in 1825. Even so, despite the opening day seeing passengers hauled by George Stephenson's Locomotion, for some years thereafter horses pulled the passenger carriages.

The opening of the Liverpool & Manchester Railway in 1830 marked the railways' coming of age in Britain, linking two major industrial areas for both freight and passengers. The country was soon gripped by railway fever, with parliamentary authorization for lines radiating out of London in all directions, as well as additional cross-country routes, such as the Grand Junction. By 1837 some 2,400 km (1,500 miles) of railway had been sanctioned, and by the end of 1843, around 3,200 km (2,000 miles) were in operation.

The railway system grew in an unplanned fashion, with private companies gaining authorization from parliament and funding the building of the lines through shareholder subscriptions. The most successful companies such as the Great Western (1835) grew by extending their network, or, like the Great Northern (1846) through amalgamation, but the competing companies had little interest in a national route map as opposed to a prospectus map of their own network designed to pull in new shareholders. Drake's maps were one attempt to resolve this difficulty, followed by Zachary Macaulay of the Railway Clearing House – established in 1842 as a way of facilitating through ticketing on lines owned by different companies – who created a national map in 1851. A series of iconic maps of Britain's railway age were also created by George Bradshaw, a Lancashire engraver, who then branched out into the publication of railway timetables in 1842 (an urgent necessity, given the number of competing railway companies).

As well as making cheap inter-city travel possible for the first time for many people, the railways also continued to provide a vital service to industry. Whereas in 1845–50 only 1.6 per cent of London's requirements had been met by coal hauled by canal or rail, in 1867 rail-transported coal exceeded that carried by sea to the capital for the first time. By 1869, the North-Eastern Railway was carrying 4.7 million tons of coal and 1.7 million tons of coke each year, dwarfing the 1.6 million tons of coal and 117,000 tons of coke it had transported twenty years earlier.

By then the railways had become established as the country's premier means of long distance transportation, a role that they would fulfil until after the Second World War, long into the age of the automobile and more than a century after Drake's first attempts to chart their course.

see more on next page >

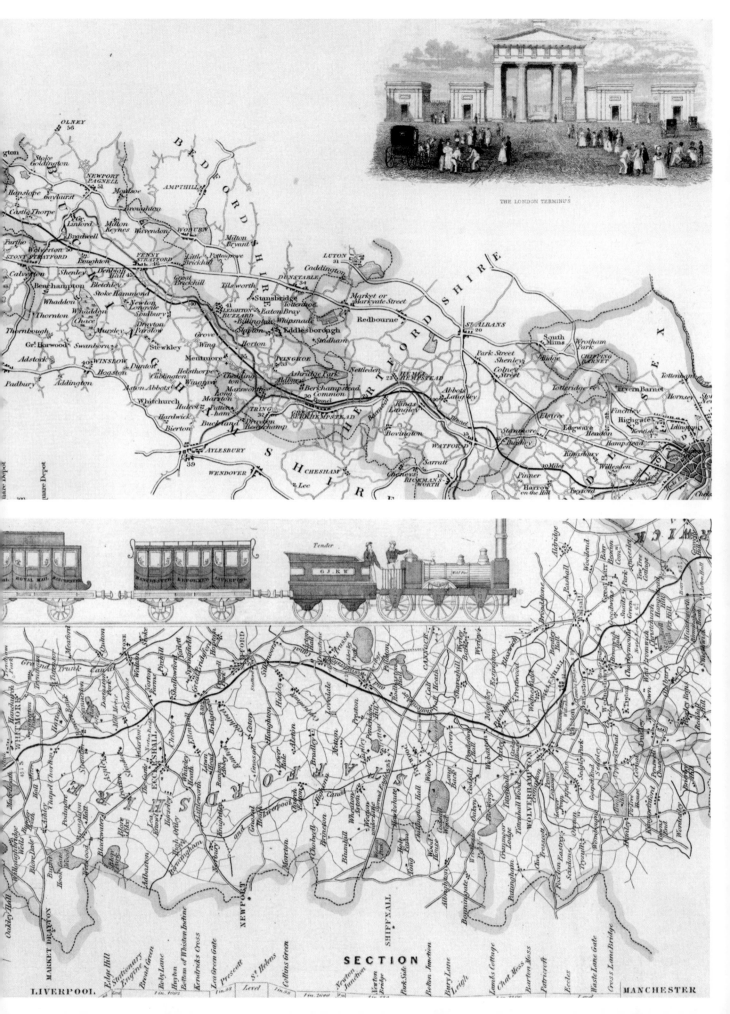

181
172
Higher
Down Park
176
175
173
170 and 171
162.
Great Halden
160
159
158
156
155
153
Old
174
Langlands
168.
Great Black Park
166
167
163
161
157
154ᵃ
154
156
120
160
165 and 132.
164.
Great New Park
161ᵃ
Long Mead
Five Acre
149
148
147
Goldcombe Farm
244
191
122
123 and 124
Sedgraft Mead
132
133.
Whealands
134.
135.
136.
161ᵃ
Lower Rutland
144
145
146.
Great Meadow
252.
Higher Six Acre
251
126.
Great Mill Mead
128.
125
120
130.
Windsor
143.
Marles
142.
Goldcombe Meadow
253.
256
254
253
Lower Six Acre
250.
Brimley Home
108
127.
Wyatt Mead
Oat Ash
106.
105.
103 and 104.
104
101.
Ham
137
138.
Horse Field
141
140
258
257.
Curleditch
259
255
254
249
248.
86.
Phoenix Thorn
107.
87.
105.
103 and 104.
94
Pomeroy House
94ᵃ
94
94ᵃ
96.
98
99
97
416
413
412.
411
410
418
417
415
416
Church
407
405.
401.
399.
308
396
395
261
262.
260
64
266
263
248.
266
269
270.
271
267
Catshay Home Mead
271
27
85.
88 and 89.
Courtney
421.
403.
400.
402.
Beer Meadow
397.
394.
84
422.
Higher Tadmoor
419 and 420.
404.
389.
392
436.
434.
435.
Middle Hill
83
423.
Clayalls
424.
Broad Field
384
385.
387.
Hodges Mead
385
389
391
390
433.
Hill Close
425

TITHE MAP OF GITTISHAM, DEVON

This map of the parish of Gittisham in Devon gives precious evidence of the layout and ownership of the agricultural land in the district in 1838, with the fields all carefully drawn, together with roads, pathways, areas of forested land and villages. Yet it was not produced in the interests of creating a modern map of Britain, but to help sweep away a medieval anachronism, the tithe payable on agricultural produce.

The obligation of landowners to pay tithes (or a tenth of their produce) to priests and temples had emerged in Biblical times and had become incorporated into Christian tradition in the early Middle Ages. In England, King Aethelwulf of Wessex confirmed the right of the church to receive them in 855 and they became an important source of ecclesiastical income, although those revenues accruing to the monasteries became alienated to secular landowners after the dissolution of the monasteries under Henry VIII in the late 1530s.

By the early nineteenth century the system had become complex and unwieldy. Precedents, exemptions and exceptions had proliferated and many – both those working the land and those owning the rights to the tithes – found it unwieldy to have to hand over actual produce once a year. Some tithes had been commuted to cash payments in the eighteenth century during the movement to consolidate isolated strips of land into more workable blocks (many of which required parliamentary approval through Enclosure Acts), but the vast majority remained in place.

In 1836 parliament passed the Tithe Commutation Act to cut through the whole rickety structure of tithes, legislating that if agreements could not be made locally, parliamentary commissioners would impose a settlement to set the levels of cash payments that would replace tithes in kind. Surveyors were despatched throughout the countryside – some of them local men, schoolmasters or petty landowners, others professionals such as Robert Pratt of Norwich who surveyed 121 districts in Norfolk and Suffolk. In all 11,785 districts were surveyed, covering 11 million ha (27.2 million acres), or three-quarters of the land in England and Wales, in a process that took until 1841 before it was largely completed. Devon, in which Gittisham parish lay, had the largest extent of land subject to tithes of any county, with around 0.65 million ha (1.6 million acres) having to be surveyed.

In each case maps were drawn of the district, as the tithes were set field-by-field and it was important to establish their correct area.

see more on next page >

Maps were accompanied by a schedule naming the landowner of each block, as well as the tithe-holder who was to receive the commuted payment. Largely the process was amicable and agreements were reached between tithe-payers and tithe-holders but in some areas, such as Kent where over 40 per cent of agreement were imposed by the government and Leicestershire where the figure was as high as 78 per cent, the commissioners had to intervene.

At much the same time as the tithe surveys were being undertaken, a new political controversy was raging that threatened the livelihood of Britain's agricultural workers. At the end of the Napoleonic Wars in 1815, the price of wheat, which had been kept high by Napoleon's Continental System – an attempt to impose a trade blockade in Britain – began to tumble. Fearing a surge of imported grain which would ruin the income of British farmers, the government passed the Importation Act, which forbade the importing of foreign wheat until the price hit 80 shillings a quarter. What was good for farmers, whose income was now protected by being able to sell domestic grain at inflated prices, was bad for the growing population of British cities and a bitter struggle broke out between the Conservatives, who supported grain tariffs and the Whigs, and particularly the Radicals, who strongly opposed protectionism.

Although a sliding scale was introduced in 1828 to allow more foreign grain in, the opposition to what became known as the Corn Laws grew. An Anti-Corn Law League was established after a meeting in Manchester in September 1838, headed by the radical campaigner Richard Cobden. The League gained large-scale support amongst the urban middle classes and used the huge funds it raised (including £100,000 in 1845 alone, in part raised by a bazaar held in Covent Garden Theatre) to produce a stream of tracts, handbills and new-sheets, all agitating for the abolition of the Corn Laws. No one was immune from their stream of vitriolic propaganda; even Lord Ashley (who, later as the Earl of Shaftesbury, became noted for his advocacy of laws to limit child labour in factories, and who helped establish the system of Ragged Schools, and so was hardly a heartless defender of the status quo) was portrayed in a wood-cut as stealing the bread from factory children.

In 1846 the Conservative government of Robert Peel caved in and a new Importation Act abolished the system of tariffs on imported grain. The price of wheat, though, did not immediately fall, as had been feared, but remained steady until the 1870s when large-scale importation from North America began, finally leading to the collapse in agricultural incomes that had long been feared. By then, the tithes had gone, too, as Britain's rural landscape began to change for good.

Map
of the
PARISH
of
GITTISHAM
Devon.

Surveyed by E. Watts — Yeovil.

1838.

We the undersigned Tithe Commissioners for England and Wales do hereby certify this to be the Map or Plan referred to in the Apportionment of the Rent charge in lieu of Tithes in the Parish of Gittisham in the County of Devon. As Witness our hands,

Wm Blamire
Tho Buller

COMBE HOUSE

GITTISHAM HILL

Westgate Hill

To Honiton

To Lyme Regis

To Sidmouth

Hare & Hounds Inn

Widbury Parish

Scale of

Chains and Links

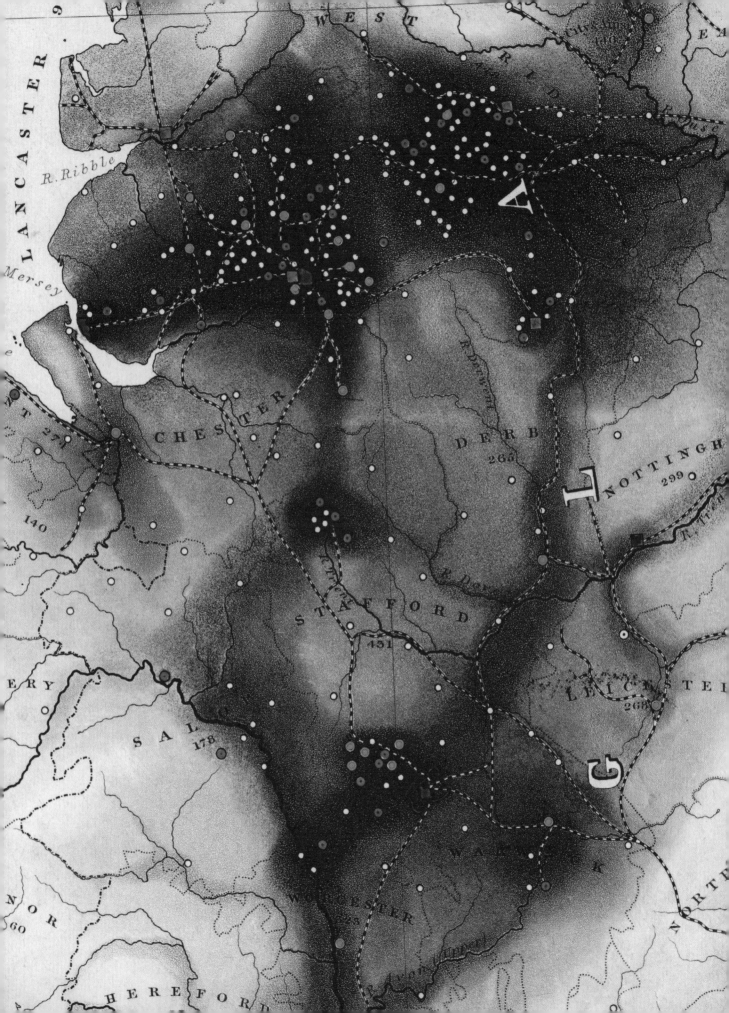

CENSUS MAP, August Petermann

August Petermann dedicated his 1849 map illustrating the 1841 census to Queen Victoria. She was the first British monarch to have access to the names of every one of her subjects (of whom the census enumerated 26,707,091) and Petermann's use of new techniques of shading and graphical presentation allowed her to view in which parts of her realm the population was most densely distributed and which of her towns and cities enjoyed the largest population.

Censuses had long been viewed with suspicion as mere pretexts for increasing the burden of taxation on the poor and with good reason. Precedents such as William I's Domesday Book in 1085–7 and the Poll Tax imposed by Richard III in 1377 and 1380 which provoked the Peasants' Revolt, seemed to confirm that monarchs were not so much interested in the numbers of men in their realm, as in the amount of money they might extract from them.

Opinions began to change in the late eighteenth century. The clear demographic and economic changes taking place as a result of the Industrial Revolution led to concerns about the government's ability to manage in the absence of concrete information on the distribution of population. Worse still, the work of the pioneer economist Thomas Malthus, led to fears that that population might simply starve. In *An Essay on the Principle of Population* in 1798 Malthus argued that although, throughout history, food production has increased, this has simply meant a corresponding increase in population. Nations, therefore, were doomed to exist on the edge of starvation and, as population tended to increase far quicker than food resources, then famine, disease and the effects of war were only Nature's way of removing the surplus in people.

To head off this apocalyptic scenario, the government needed to know how many heads it had to feed. The solution came from an unlikely source. John Rickman, a Clerk in the House of Commons proposed in 1800 that the ministers of every parish in the country send in returns on the birth, marriages and deaths in their district for the last ten years, then a more detailed enumeration of just three or four parishes (at a cost

see more on next page >

Rickman estimated, of a mere £740) could be used to calculate a precise tally of the nation's population.

Rickman's idea was taken up in a modified form and he helped draft a parliamentary Act which enabled the carrying out of a census on 10 March 1801. As well as ministers, town clerks and overseers of the poor were seconded to help ensure that every household filled a set of four questions (including on occupation). The process was new, and there was still understandable suspicion about its motives. As a result, the returns were partial in some cases (and for Monmouthshire almost nonexistent), and so the final enumeration contained a good deal of extrapolation. Yet the total of 10,901,236 for England, Scotland and Wales was the most accurate snapshot of the nation's population which anyone had yet achieved. Revealingly, many areas which would be at the heart of city-centres a century later, were still semi-rural. Paddington (now in central London) was a farming community of 1,881 and Fulham to the city's southwest had 450 people working in agriculture.

The census was repeated every ten years thereafter (save in 1941 when the Second World War made carrying it out impossible). In 1837 the establishment of the Registrar General of Births, Marriage and Deaths for England and Wales created an apparatus which could carry out the census far more efficiently than ad hoc local boards and by the time of Rickman's death in 1840, the census had become an established national ritual.

The 1841 census was the first to ask the names of all persons present in a household on census day. It found that the population had risen to nearly 27 million, close to triple the level of 40 years previously. Petermann, a German cartographer who had settled in London in 1847 and established a thriving business producing atlases, identified a wonderful opportunity in the wealth of statistics the census made available. He had pioneered new techniques in using mapping to display statistical information and employed them to full effect on the census map: areas of dense population are shaded more heavily, while larger rings indicate cities with a greater population and graphs are used to show the increase in population of the country over time. In gifting this map to Queen Victoria, Petermann had given new meaning to the sentiment that she was 'sovereign of all she surveyed'.

DIAGRAM
showing
by a Scale of Per Centage
the
COMPARATIVE PROGRESS of POPULATION
in all the
CITIES & TOWNS of GREAT BRITAIN
containing
100,000 inhabitants and upwards
from 1801 to 1841.

Glasgow, City and Suburbs.
Liverpool, with Toxteth Park.
Manchester, Salford and Suburbs.
Leeds.
Birmingham, and Suburbs.
Edinburgh, City including North and South Leith.
London.
Bristol, with Barton Regis.

Table showing the Average Density of Population
or the
NUMBER OF SOULS TO 1 ENGLISH (STATUTE) SQUARE MILE
in each County etc. of the British Isles.
Arranged according to the Amount of Density.

E = England · W = Wales · S = Scotland · I = Ireland · V = Isles in the British Sea.

General Remarks.

DIAGRAM
showing
by a Scale of Per Centage
the
COMPARATIVE PROGRESS OF POPULATION
in
ENGLAND, SCOTLAND AND IRELAND
from 1571 to 1841.

DIAGRAM
showing
by a Scale of Per Centage
the
COMPARATIVE PROGRESS OF POPULATION
in
ENGLAND, WALES, SCOTLAND & IRELAND

Synoptical Table
of the
Number and Total Population of all Towns containing
10,000 inhabitants and upwards.

Summary Account
of the
TOTAL POPULATION and AREA of the UNITED KINGDOM.

To Her Most Excellent Majesty
QUEEN VICTORIA
THIS MAP OF THE
BRITISH ISLES,
ELUCIDATING THE DISTRIBUTION OF THE
POPULATION,
BASED ON THE CENSUS OF 1841;
Compiled and drawn by
AUGUSTUS PETERMANN F.R.G.S.
Honorary Member of the Geographical Society of Berlin.

By Her Majesty's gracious permission most humbly dedicated By the Author.

Explanation.

THE SHADING *exhibits the various degrees of density in every part of the United Kingdom.*

THE FIGURES *denote the average amount of density of the population in each County, namely the number of souls to 1 English (Statute) Square mile.*

THE COLOURED SPOTS *indicate the position of all Towns containing 5000 inhabitants and upwards, according to the following arrangements.*

Towns of 5000 and 10000 inhabitants and upwards.
From 10000 to 20000 inhabitants
From 20000 to 50000 inhabitants
From 50000 to 100000 inhabitants
From 100000 to 200000 inhabitants

Railways completed.

Scale of $\frac{1}{1.600.000}$ of Nature = about 25 Miles to 1 Inch.
English Miles.

GUERNSEY
and adjacent islands.

JERSEY with

GROUND FLOOR

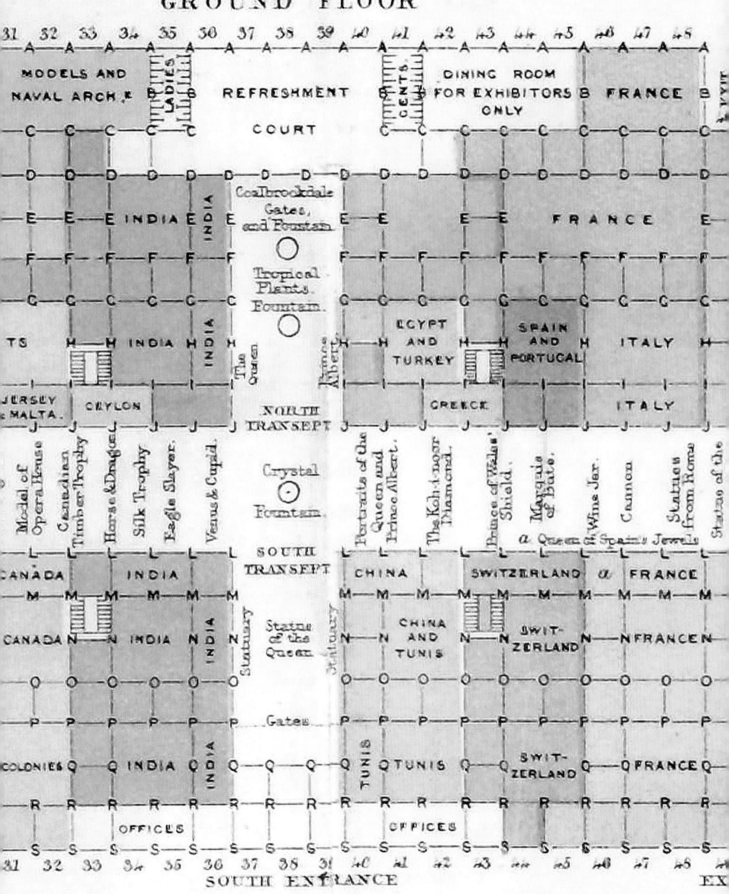

MODELS AND NAVAL ARCH.ᵗ

LADIES

REFRESHMENT COURT

GENTS

DINING ROOM FOR EXHIBITORS ONLY

FRANCE

EXIT

INDIA INDIA INDIA

Coalbrookdale Gates, and Fountain

Tropical Plants. Fountain.

The Queen

Prince Albert.

FRANCE

EGYPT AND TURKEY

SPAIN AND PORTUGAL

ITALY

JERSEY & MALTA.

CEYLON

INDIA INDIA

NORTH TRANSEPT

GREECE

ITALY

Model of Opera House.

Canadian Timber Trophy

Horse & Dragon.

Silk Trophy

Eagle Slayer.

Venus & Cupid.

Crystal Fountain.

Portraits of the Queen and Prince Albert.

The Koh-i-noor Diamond.

Prince of Wales' Shield.

Marquis of Bute.

Wine Jar.

Cannon

Statues from Rome

Statue of the

Gates

Statuary

Statuary

SOUTH TRANSEPT

CHINA

SWITZERLAND

a. Queen of Spain's Jewels

a FRANCE

CANADA

INDIA

Statue of the Queen

CHINA AND TUNIS

SWIT-ZERLAND

CANADA

INDIA

INDIA

INDIA

FRANCE

TUNIS

COLONIES

INDIA

INDIA

TUNIS

SWIT-ZERLAND

FRANCE

OFFICES

OFFICES

SOUTH ENTRANCE

EX

Scale of Feet   100   150   200 Feet

GREAT EXHIBITION PLAN

The ground-plan of the Great Exhibition of 1851 is a symbol of Victorian Britain at the height of its self-confidence. Row upon row of stands displayed the finest examples of the nation's industrial prowess, all encased within an enormous palace of glass, which was the marvel of the world.

By 1851, Britain had no equal, diplomatically, politically or economically. Possessed of an empire which straddled the globe, from Australia to India to Canada, its industrial might was one which no other country – at least for several decades – could match. By the mid-nineteenth century, British mills produced around half the world's output of cotton textiles and British furnaces about half of its steel. Raw cotton consumption alone had grown from 1.2 million kg (2.6 million pounds) in 1760 to 282 million kg (621 million pounds) in 1850, as the textile mills of northwest England eagerly consumed all that Britain's growing shipping fleet could bring them. Britain's cities were bursting too, while London more than doubled in population from 0.94 million in 1801 to 2.32 million half a century later, cities in the north began to rival it for the first time, with Manchester expanding from being practically a village to a bustling, teeming urban hub of 180,000 people by 1831.

Demand called forth innovation and Britain's inventiveness (from steam engines to trains to blast furnaces) was the envy of other nations. All that was lacking was a showcase in which Britain could display its industrial supremacy for all to see. The determination of the Royal Society for the Encouragement of Arts to create one, coupled with the energy of Prince Albert, Queen Victoria's husband, who had found the role of Prince Consort something of a disappointment, provided the genesis of the Great Exhibition.

By 1850, proposals were complete, and the architect Joseph Paxton was drafted in to build a hall fit for an exhibition on an imperial scale. True to his background as a designer of greenhouses for the Duke of Devonshire, Paxton came up with an audacious plan. Essentially a vast greenhouse, the Crystal Palace – as it came to be known – was 564 m (1,851 ft long, in a reference to the year) and contained around 93,000 sq. m (1 million sq. ft) of glass and 38.5 km (24 miles) of guttering.

On the chosen site in Hyde Park, legions of workmen hurried to make everything ready and then, in October 1850 the exhibits began to arrive. On the opening day, 1 May 1851, a huge crowd of 300,000 people gathered to see the Duke of Wellington – who was celebrating his eighty-second birthday – together with a glittering array of foreign ambassadors, archbishops and generals, enter this temple to British industry.

Inside they and the 42,000 people a day who paid the shilling entrance fee were treated to an astonishing array of exhibits, some mundane (but still impressive) examples of Britain's industrial power, others frankly outlandish. Visitors could see displays of machinery which ran the whole gamut of textile production from raw cotton to finished cloth, but they could also gawp at Frederick Bakewell's primitive fax machine, an automatic voting machine and an alarm clock which caused the unfortunate owner's mattress to tilt, pitching him out of bed at waking-up time. Those prepared to queue could also catch sight of the massive Koh-i-Noor diamond, at 186 carats the world's largest, which had only recently arrived in Britain. Among Queen Victoria's favourite exhibits was a piano adapted so a quartet could play on it.

Although the vast majority of the exhibition space was devoted to British wares, other nations were allowed to display their goods. The French brought fabrics and a prototype submarine, while the Netherlands saw fit to include a gargantuan 1.2 m by 0.6 m (4 ft by 2 ft) pie on its stand. Politics intruded, too, as the smaller German states such as Hamburg and Bremen refused to join Prussia's stand, a mirror of the situation in Germany where they were afraid the Prussians were about to snuff out their independence.

The Americans were mocked for the stuffed squirrel and the huge eagle they suspended from the ceiling to hover over their stand. Yet the range of their exhibits surprised many and their range of unpickable locks were a sensation, and a sign that Britain, content for the most part to rest on its laurels, would before too long be displaced by the United States and Germany as the world's industrial engines. In October the exhibition closed, and the whole structure was dismantled and transported to Sydenham, in south London (where it stood until destroyed by a fire in 1936). The crowds who had flocked its aisles, gazing at the wonders and complaining at the coldness of the coffee and the price of the ice cream, dispersed. By the time the Festival of Britain celebrated the centenary of the Great Exhibition in 1951, it was amid an atmosphere of nostalgia for imperial decline far removed from the spirit of its Victorian predecessor.

see more on next page >

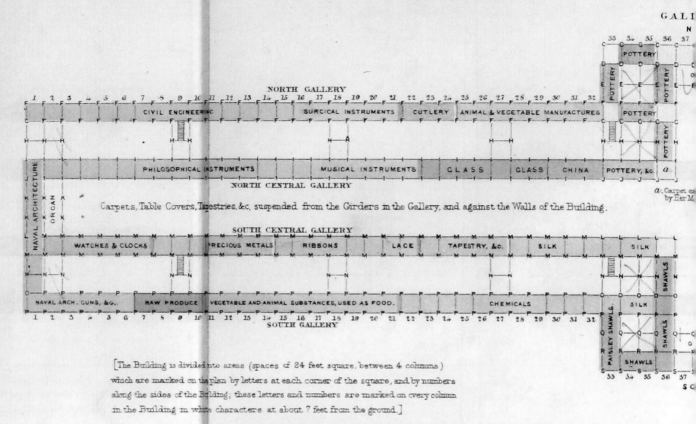

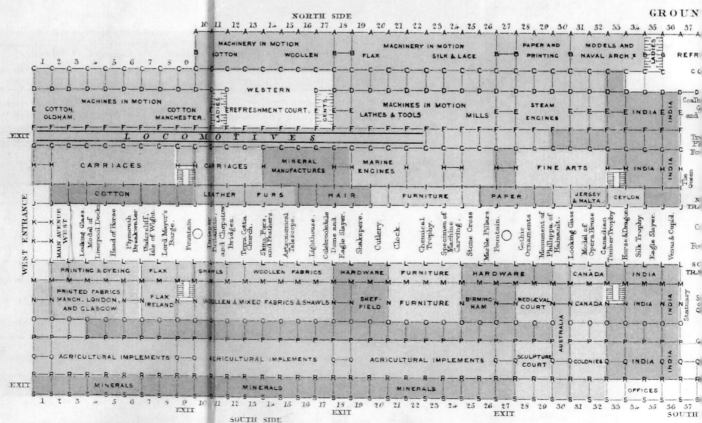

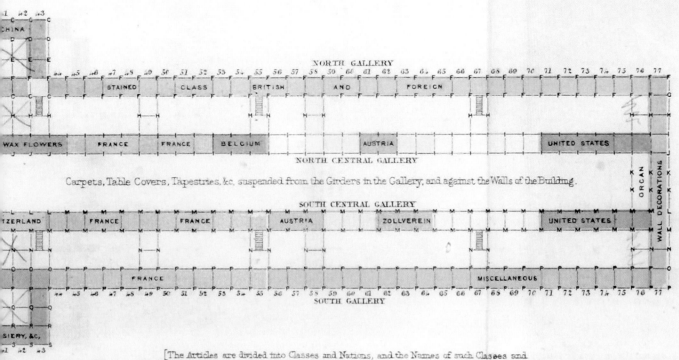

Carpets, Table Covers, Tapestries, &c. suspended from the Girders in the Gallery, and against the Walls of the Building.

[The Articles are divided into Classes and Nations, and the Names of such Classes and Nations are given on the Plan, and marked upon the iron girders of the Building.]

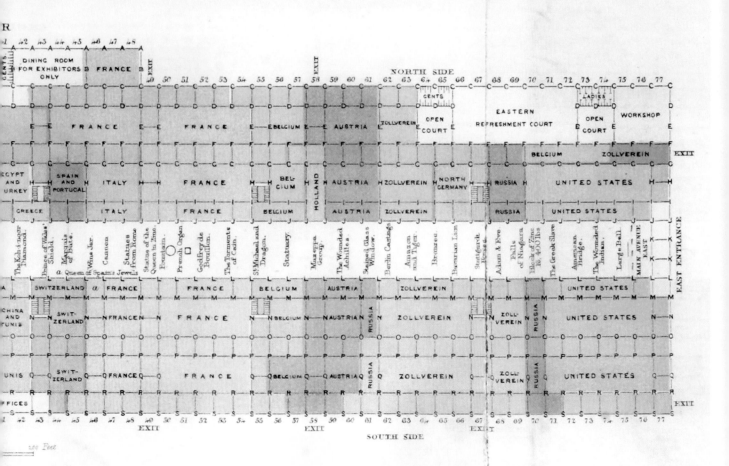

CHOLERA MAP, John Snow

The map which London doctor John Snow produced of an outbreak of cholera in London in 1854–5 played a key role in overturning long-held beliefs about how the disease spreads. Each of the black bars records a case of the disease in that outbreak. Cholera had stalked Victorian London for decades, striking fear into a city which understood nothing of how it was caused and was, until Snow's work, powerless to prevent its spread.

Cholera, a deadly disease which appeared to have originated in India, struck European countries with increasing severity after 1830. The prevailing medical orthodoxy was that it was an airborne disease spread by 'miasma', the poisonous vapours found in abundance in the squalid and overcrowded cities of the time. The stench which drifted from the squalid waters of the Thames was considered quite enough on its own to contaminate the air and spread any number of fatal maladies. Dr John Snow, a London anaesthetist, was uneasy with this simple equation, considering that the disease-producing vector must be ingested by the victims in some way.

In 1854 a new cholera epidemic struck London and Soho was hit particularly hard, with 500 deaths in just five days. The area had piped water for only two hours a day and so many of its residents depended on water from pumps. Snow undertook a detailed investigation and found that the level of deaths from those who took their water from one particular pump, at Broad Street, was far higher than among those who relied on other water sources – of 137 people he could find who had definitely drunk water from the pump, 80 died.

Snow persuaded the members of the Parish Board of St James to remove the pump handle, so rendering it useless. Before long, deaths from cholera in the immediate area fell dramatically. More crucially, he charted his findings on a commercial map, marking a series of black bars against addresses to mark the number of fatalities. He was not the first to map the spread of a disease – as early as 1832 A. Brigham had drawn a 'Chart Showing the Progress of the Spasmodic Cholera', nor the only one to map the 1854 Soho outbreak (Edmund Cooper working for the Metropolitan Commissioner of Sewers had done that), but his plotting of the water pumps as a means of diagnosing the cause of the disease was revolutionary. A further refinement in a second version of the map in which he added a boundary line around those areas where the Broad Street Pump was the closest water source was an even more vivid demonstration of his point. Contaminated water was the source of cholera, and the scientific mapping of medical conditions was a phenomenon that was here to stay.

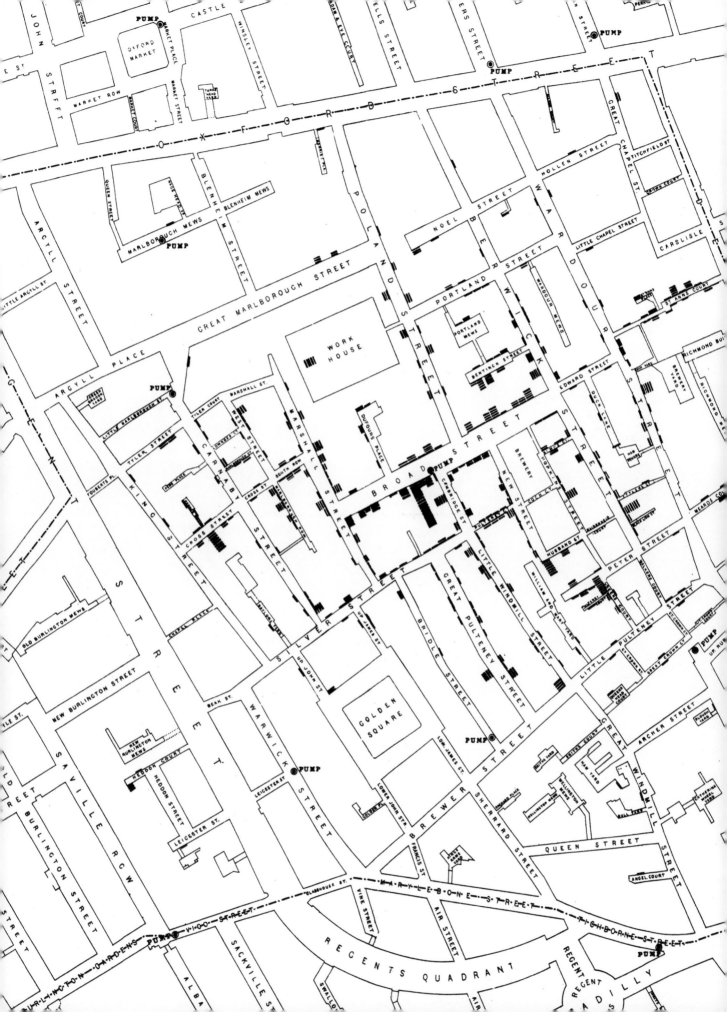

WEATHER CHART, MARCH 31, 1875.

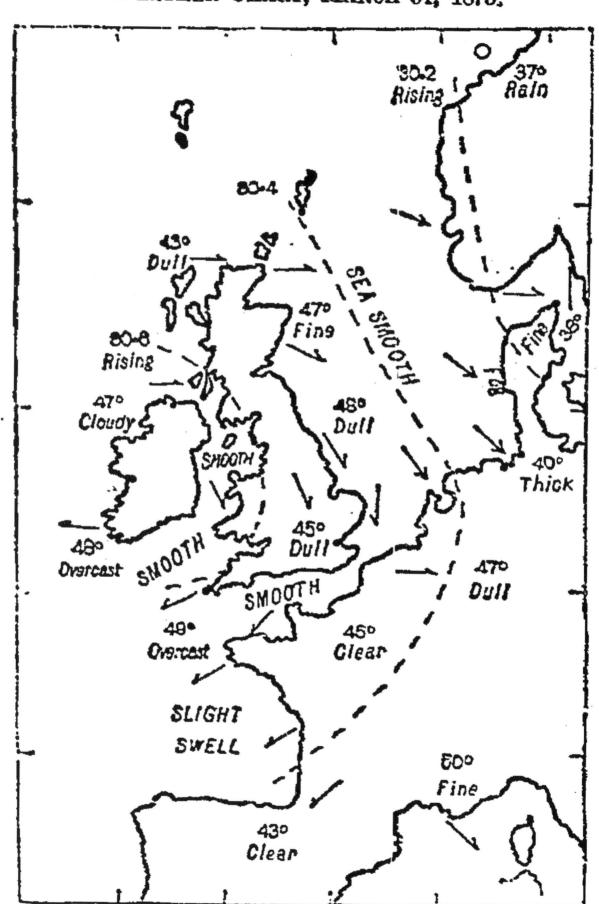

THE FIRST WEATHER MAP

On 1 April 1875, The Times published the first weather map to appear in a British newspaper. Drawn up by Francis Galton, a polymath with a fascination for meteorology, it gave a bare description of temperatures and atmospheric conditions over the preceding twenty-four hours, but provided a new outlet for the perennial British passion about the weather.

Two cousins and two suicides formed the backdrop to the April 1875 weather map. Galton was the first cousin of Charles Darwin, whose pioneering voyage on the *Beagle* to the South Atlantic in 1831–6 had led him to formulate his theory of evolution that rocked the Victorian scientific and religious establishments. Darwin's captain on that voyage was Robert FitzRoy, who had been persuaded that the young naturalist might make a suitable companion to dispel the gloom that had caused his predecessor on the *Beagle*, Pringle Stokes, to blow his own brains out (having written in his diary shortly before that 'The weather was that in which "the soul of man dies in him".')

For FitzRoy, the weather became the source not of depression but of determination and after his return to Britain he dedicated himself to finding a solution to warning mariners of the storms which so often caused shipwrecks (leading to the loss of 7,200 lives between 1855 and 1860). In 1854, he was appointed head of the new Meteorological Office, which was intended as a repository of maritime charts, but whose mission he moulded to encompass the issuing of storm warnings. Using the recent discovery of the electrical telegraph, FitzRoy established a network of telegraph clerks along the coast and in February 1861 issued his first storm warning, dividing the United Kingdom up into six geographic areas to make the preparation of the reports easier. Soon thereafter, he advanced to issuing weather forecasts, which *The Times* began printing on 1 August 1861 (when it predicted a cool day at 61 °F in Liverpool, with a light southwesterly breeze).

Many were thrilled, including the organizers of large events who could now take precautions against predicted rain, and gamblers who could place bets on the horses with an inkling of conditions expected on the racetracks. Many, though, were cynical – an MP had been greeted with howls of derision in the House of Commons when he predicted in 1854 that before long the weather in London could be known twenty-four hours in advance – and fishermen grumbled that incorrect forecasts from FitzRoy had lost them a day's catch. More importantly, the doyens of the Royal Society took ill the pretensions of an outsider such as FitzRoy and lost no chance to snipe at him.

FitzRoy became depressed and also felt guilty about his association with the spread of the theory of evolution. He was religiously very conservative and his own account of the *Beagle* voyage speculated that the reason for the extinction of megafauna such as the mastodon and dinosaurs was simply that they had been too large to fit through the doors of Noah's Ark. The criticism of his beloved weather forecasting-service must have played a role, too, and on 30 April 1865 he locked himself in his dressing-room and, just as his predecessor on the *Beagle* had done, he committed suicide.

Among those who had criticized FitzRoy was Francis Galton. A prodigious intellect – who by repute had mastered the Ancient Greek alphabet when eighteen months old – Galton's breadth of scholarship was breathtaking. His work on statistics led him to an unhealthy brush with eugenics (the notion that the genetically most promising humans should be encouraged to breed more) and he devised a scientific classification of fingerprints to help in the solution of crimes. His passion, though, was meteorology – he discovered the anticyclone in 1861, the atmospheric phenomenon by which winds spiral around a point of high pressure.

Galton was commissioned to write a report in the aftermath of FitzRoy's death examining the benefits of weather forecasting. His forensic mind damned the amateurish prognostications of his competitor. There was no method, no formula in them (Galton in contrast had once devised a formula to create the ultimate cup of tea). As a result, the storm warnings were suspended in 1866, though restored the next year as the British public (including its mariners) had become too attached to them.

The main weather forecasts were not resumed until 1879, but Galton made some amends for his treatment of FitzRoy by continuing to argue for the cause of meteorology and in 1875 he produced his first weather map for *The Times*. Few of its readers, though, will have known of the storms which lay behind its production.

ELECTRICITY SUPPLY MAP OF LONDON

The map charts how light came to London. With its different coloured zones marking out the franchises granted to electricity companies in the capital in the 1880s, it demonstrates how technological advances originally conceived to benefit industry were finally bringing benefits to Victorian Britain's consumers.

Mid-Victorian London – as all of Britain's cities – was a lively, confusing and often dangerous place. The city's population had reached 1.6 million by 1831 and would balloon to 2.8 million by 1861 and 5 million by the end of the century. The ever-present crime that plagued Londoners was in part stemmed by the establishment of the Metropolitan Police in 1829 (and the prospect of hanging for relatively petty crimes, a gruesome sentence carried out in public until 1868), but disease continued to exact a terrible toll. London's water was infected with a lively collection of dangerous organisms and an outbreak of cholera in 1832 (the first of several), claimed the lives of 800 people. So loathsome was the pollution in the River Thames that in 1858 the stench became overpowering and the 'Great Stink' forced disgusted MPs to legislate that the Metropolitan Board of Works be responsible for cleaning up the river. The Board's chief engineer Joseph Bazalgette finally complete a system of proper sewerage for London in 1875.

All of this took place in comparative darkness. For centuries rush-lights, oil lamps and tallow-candles (or for the better-off, beeswax candles) had provided a fitful halo of illumination during the night-time hours. In 1807 gas lighting appeared, first demonstrated by the German entrepreneur Frederick Albert Winsor on Pall Mall in honour of George III's birthday. The novelty soon took off, and by 1823 over 320 km (200 miles) of London's streets were lit by gas.

The system had its drawbacks, however. The light was weak – providing more oases in the gloom than well-illuminated streets. Each light also had to be lit and extinguished every day at dusk and dawn, an expensive process as the network spread. When used domestically, gas lights also had an alarming tendency to explode or to reduce domestic drawing rooms to soot-choked, oxygen-starved parlours. The invention of an electric arc lamp by Frederick Holmes in 1846 seemed to offer a solution, but the practical difficulties of implementing it on a large scale held back development until the installation of a variant (the Yablochkov candle) in Paris in the 1870s.

The invention of a functional incandescent lamp by Joseph Swan in 1878 (more or less at the same time as the American inventor Thomas Alva Edison) showed that electricity's time had come. Already the Gaiety Theatre in London had been lit (albeit by arc lights), and in June 1881 the House of Commons followed, with the Savoy Theatre soon after.

Shares in fledgling electricity companies soared, as those in the gas businesses faltered. The gas industry was so concerned it appointed a committee to investigate, which reported back that 'we are quite satisfied that the electric light can never be applied indoors without the production of an offensive smell which undoubtedly causes headaches.' Undeterred (and correctly), the investing public ignored these predictions and, with £1.5 million of new capital behind them, the electricity companies pressed for legislation to permit them to lay cables underground.

It took three years of debate before the Electric Lighting Act of 1882 was passed, which allowed local authorities, companies and private individuals to generate electricity and dig into public streets to lay their cables. A pioneer generating station was established by Edison at Holborn Viaduct in January 1882 but it, and several like it, had closed down by 1884 because they insisted on matching the price of gas (as Edison's company did in the United States, where gas was more expensive, and so the promise was cheaper to uphold).

A generating station clung on at Victoria Station however, and smaller scale projects supplying smart residential terraces in London, and larger areas in towns such as Brighton, proved that electricity was a viable source of power. In industry there was less doubt and, starting with Bradford in 1892, soon almost all of Britain's trams were electric, while the London Underground's first electric train started operating in 1894. Factories, mines and shipyards also soon realized the benefits of electricity, so that by 1895 38 gigawatt hours were being generated a year.

In 1888, a new Electricity Act was passed, doubling the length of franchises granted to electricity companies to forty-two years. This and technical advances, such as the development of the steam turbine by Sir Charles Parsons in 1884, emboldened the electricity companies to fresh investment, and soon the whole of the capital was covered by electricity franchises. Night-time would never be the same again.

see more on next page >

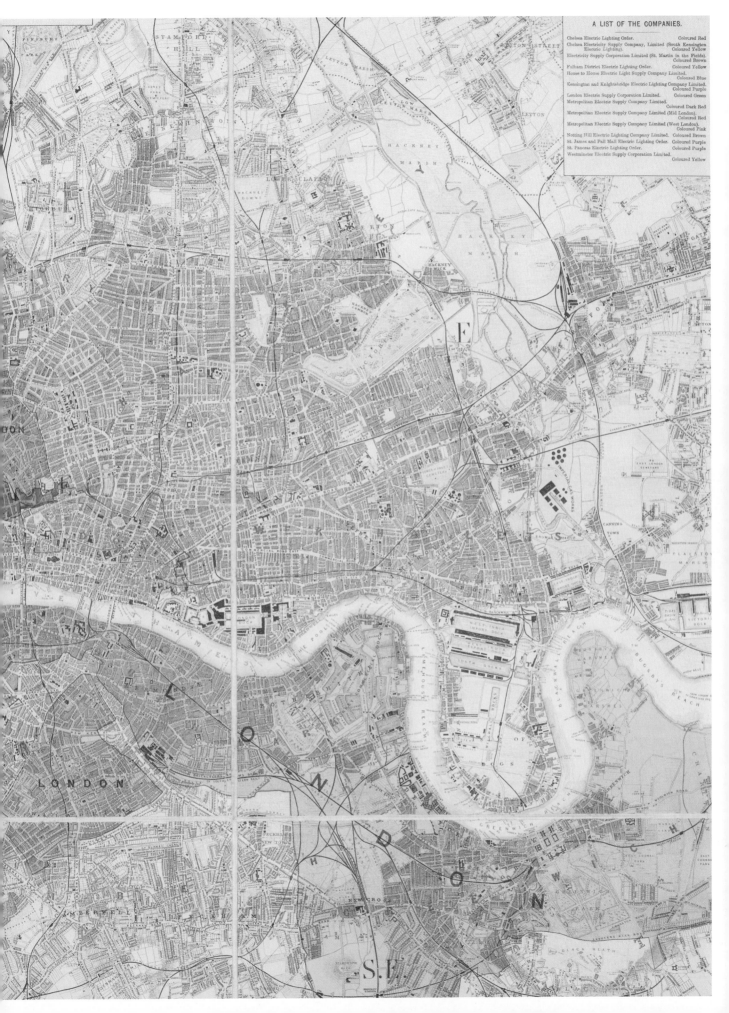

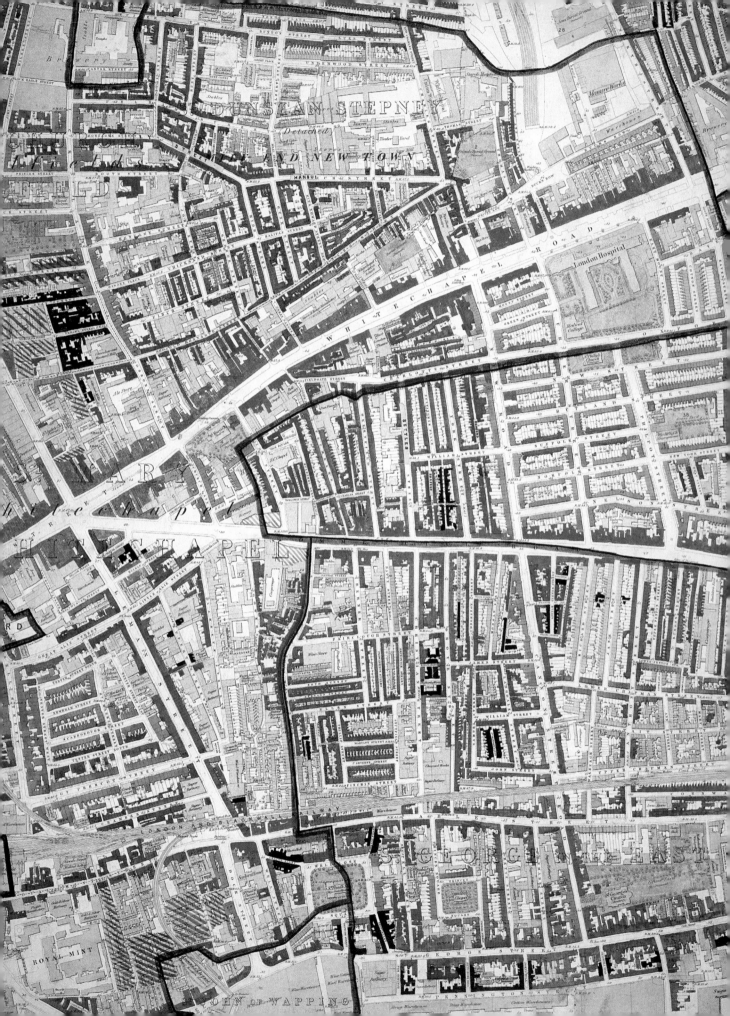

POVERTY MAP OF LONDON,
Charles Booth

The pell-mell growth of Victorian London's population brought with it severe social costs. For every industrial fortune made, there were thousands of broken lives lived in squalor and hopelessness. While some, such as the philanthropists Joseph Rowntree (in education), George Peabody (in housing) and Elizabeth Fry (in prisons) tried to improve the lot of the poor, others sought to document it. Charles Booth's maps of poverty in Victorian London were the most developed example of this, and helped crystallize unease about the plight of the urban destitute.

The population of Britain's cities grew explosively in the mid-nineteenth century, with London's rising from 1.6 million in 1831 to 3.9 million by the 1881 census. These sinkholes of urban deprivation exercised legions of Victorian reformers, who pressed for health reforms, education for the poor or the wholesale renovation of the cities' blighted cores. Yet poverty wore a very personal face and scientific tools for its diagnosis and remedy were frustratingly slight.

All this changed with the work of Charles Booth, a social scientist of astonishing dedication who devoted seventeen years to a detailed survey of the plight of London's poor. Beginning in 1886, Booth and his team of enquirers, largely drawn from the ranks of the School Board visitors established by the 1868 Endowed Schools Act, visited thousands of streets and countless homes, and trawled through census returns in an effort to establish the exact level of poverty in the metropolis. The first volume (of what would become seventeen by the work's completion in 1903) of *The Life and Labour of the People in London* determined that 35 per cent of the East End – the subject of the initial enquiry – lived in a state of abject poverty.

This was shocking enough, but what gave Booth's work added bite were the colour-coded maps which he included in the second (1891) and subsequent volumes. Every street in the capital was shaded in one of seven colours from yellow, denoting the prosperous quarters of the upper classes, through reds and blue to a sombre black signifying the haunts of the 'lowest class of occasional labourers, loafers and semi-criminals'. The intermingling of rich and poor, but above all the vast swathes of London where the majority lived below Booth's chosen poverty line of 18 shillings a week, provided a powerful visual representation of the challenges that faced reformers.

In themselves Booth's maps changed little, but they did provide ammunition for those clamouring for change. Already in 1875, the Public Health Act had made it compulsory for new residential dwellings to be connected to running water and the sewage system. Local authorities were required to have medical and sanitary inspectors to enforce existing legislation on public hygiene. A further Act in 1904 reinforced these measures and made it illegal for landlords to rent out lodgings which were deemed sub-standard by the new legislation. In Booth's hands, the map had become a powerful tool of social reform.

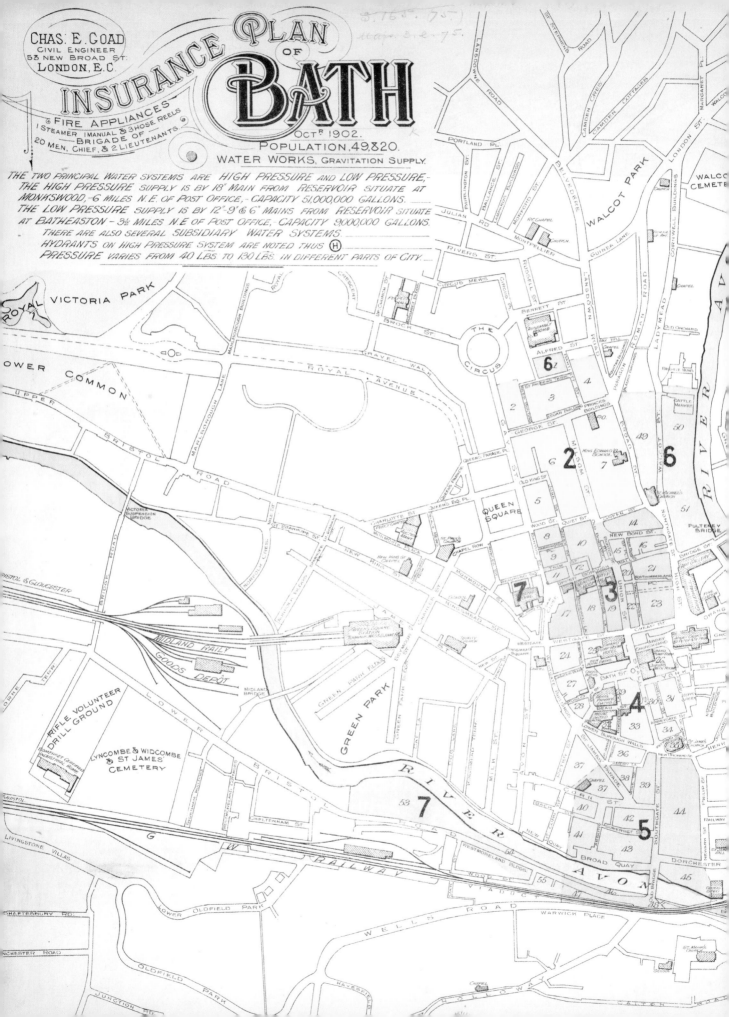

FIRE INSURANCE MAP OF BATH, Charles Goad

Fire has been an ever present hazard in cities from ancient times, but it took until the nineteenth century for mapping to be harnessed to analyse the dangers. This Fire Insurance Map of Bath produced by Charles Goad in 1901 divides the town into colour-coded sections, while more detailed sheets identified blocks and buildings at particular risk of fire.

Goad was a cartographer and civil engineer, who began producing fire insurance maps in Montreal in Canada in 1875. He returned to his native England ten years later to answer the growing demand of the insurance industry for the latest techniques to reduce their risks. The problem was the almost uncontainable growth in the cities, even outside London: Liverpool grew from 80,000 inhabitants in 1801 to over 700,000 in 1901, while Bath, a minnow at 7,000 inhabitants in the late eighteenth century had swollen to 57,000 by 1831 (accompanied by a divergence between the Georgian grandeur of the Royal Crescent, constructed in 1767–74 and a growing number of urban slums).

The increase in inner-city industries meant that there were warehouses filled with chemicals and combustible wares such as textiles, side-by-side with kilns, ovens and building materials. It required the tiniest spark to set all this ablaze with disastrous consequences. London provided a stark warning of what could happen, when a fire at a baker's shop in Pudding Lane in September 1666 began the Great Fire of London, which burnt for four days, destroying some 13,000 houses and rendering 100,000 people homeless.

Many individuals were ruined, and the government was slow to provide compensation, so business-owners turned to an alternative means of providing against any repetition of the catastrophe. In 1680 Nicholas Barbon established the Fire Office, which in exchange for the payment of a small premium, would guarantee policy-holders recompense in the event of a fire. Soon, competitors copied Barbon, and the Friendly Society, the Hand in Hand Fire Office and (from 1710) the Sun Fire Office were all jostling for business. The model also spread into the provinces and by 1767 Bath had its own fire office, too.

Before long, the fire office companies realized that a better way to avoid paying out large amounts of compensation was to prevent the buildings burning down in the first place. They established their own fire services, with engines housed in stations located near the buildings they were insuring (which were indicated by fire marks belonging to a particular fire office) In the event of a blaze, the fire service of one fire office would not come to the rescue of a building covered by another.

The chaos of conflicting companies may have been commercially profitable, but it did little to ensure the safety of Britain's cities (and created the perverse situation where a system intended to pool risk did not do so when fire services stood by and watched buildings burn). In 1833 the London fire offices finally agreed to co-operate and established the London Fire Brigade, while municipal fire services, operated by the local authority rather than private companies, began to spread (notably to Manchester where one was established in 1832).

A series of high-profile fires in London, beginning with the destruction of the old House of Commons building in 1834 and a massive blaze at Tooley Street in Southwark, where a warehouse packed with hemp caught fire in 1861, leading to £2 million of insurance claims, showed that the old fire office model was outdated. In 1865 the Metropolitan Fire Brigade Act was passed and in 1866 a new London Metropolitan Fire Brigade began to operate. One by one the regional fire offices began to close their fire services, with the last one, in Norwich, disbanding in 1929.

The appearance of a more comprehensive fire service, though, did not diminish the need for insurance and tools which identify high-risk premises (which needed to pay higher premiums for protection) were eagerly sought by insurers. So when Charles Goad began to produce his high-quality, intensely detailed maps of fire risks, he found a ready market. So popular were they that maps were produced for towns across England, Wales, Northern Ireland (including Belfast and Londonderry) and Scotland (including Edinburgh, Glasgow and Dundee). Although the insurance companies did not now need to despatch their own engines to fight fires, they could see at a glance where they were most likely to break out, and charge their clients accordingly.

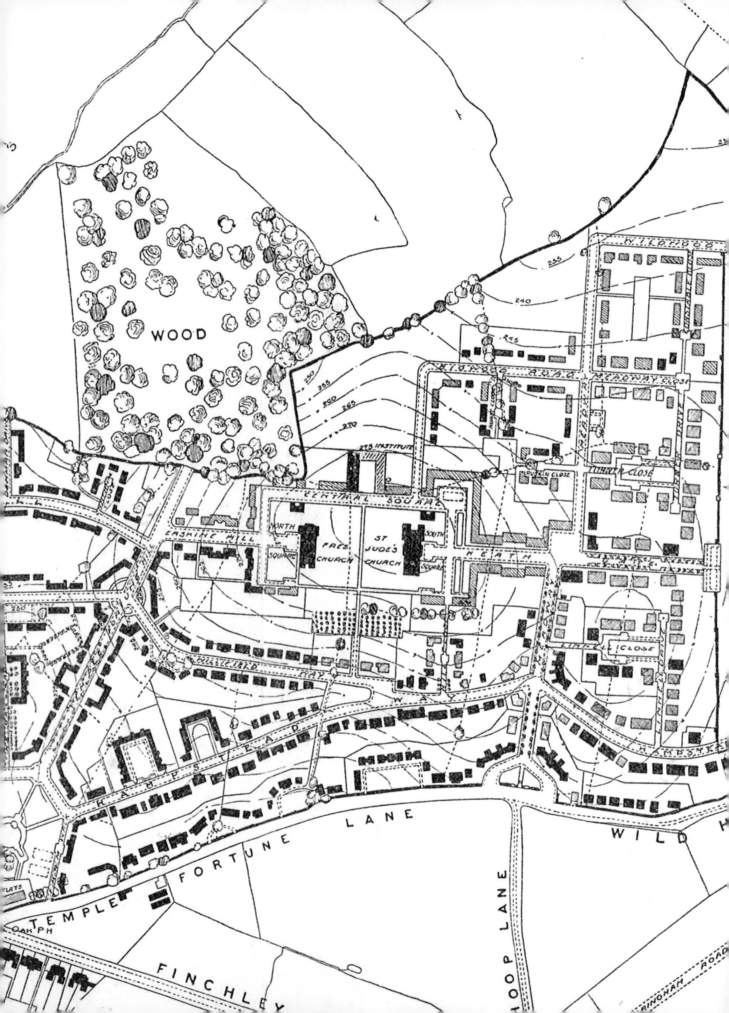

HAMPSTEAD GARDEN SUBURB PLAN

A complex pattern of closes, cul-de-sacs and squares fill this 1911 plan of Hampstead Garden Suburb, in a stark difference to the dense chaos of London's traditional street pattern or the monolithic grids of late Victorian suburban speculators. It was one of a wave of new suburbs and garden cities which sought to express philanthropists' hopes for a better life for urban workers through improved housing conditions and a more liveable environment.

Late Victorian London was a city transformed – it had burst out of its traditional medieval core in a dizzying expansion that had seen the population soar from 939,000 people in 1801 to over 4.9 million a century later. Yet it was not necessarily a city improved. Much of the increased population had been housed by subdividing existing stock or by jerry-building new estates and tenement blocks which teemed with vermin as much as they did with people. The cholera epidemics from the 1830s had stimulated concern for the conditions in which the poor lived – though in part for fear that the contagion of such diseases might touch the lives of the better off – and social commentators and novelists such as Charles Dickens painted vivid portraits of the squalor in which the masses were forced to subsist. Andrew Mearns's description in his 1883 pamphlet *The Bitter Cry of Outcast London* of 'courts reeking with poisonous and malodorous gases arising from accumulation of sewage and refuse scattered in all directions and often flowing beneath your feet' was typical of the horrified reaction of Victorians with a conscience.

The outcry for better conditions at first met with a muted official response. The 1848 Public Health Act had specifically excluded metropolitan London from its provisions, in part because the fragmented nature of London's government made concerted measures there difficult. Only in 1875 did the Artisans and Labourers Dwelling Act make action possible, although this was largely confined to slum clearances, where those displaced were often rehoused in tightly packed barracks-like complexes.

Instead it fell to private philanthropists to finance, plan, lobby and develop a new philosophy for urban living. The American banker, George Peabody, who moved to London in 1837, had asked the social campaigner Lord Shaftesbury what he could do to help improve the appalling conditions he witnessed there. As a result, in 1862 he set up the Peabody Fund to finance low-rent housing for the labouring poor and in February 1864 built the first of many Peabody developments, in Spitalfields, just east of the City of London.

Industrialists were amongst the earliest to understand the benefits of better conditions for their workforce and William Hesketh Lever's Port Sunlight (on the Wirral in Merseyside), begun in 1888, acted as the model for subsequent model villages. By then, a strong movement had emerged promoting low-density housing for the less well-off within large cities (or if that were not possible, by building completely new towns). Ebenezer Howard, who in 1902 published *Garden Cities of Tomorrow*, was among the most vigorous proponents of this trend, which came to fruition with his building of Letchworth Garden City from 1903.

Henrietta Barnett, and her husband Canon Augustus Barnett, were longstanding supporters of efforts to provide more dignified conditions for the urban poor. In 1877, they had built a model block in Whitechapel, in part financed by selling her family jewels. When a scheme was put forward in 1902 to extend London Underground's Northern Line to Golders Green, with the uncontrolled speculative building that would likely accompany it, Henrietta Barnett stepped in with an alternative proposal.

Barnett laid out a vision inspired by all the work of preceding housing philanthropists to create a suburb where houses were not to be built in soulless uniform lines, where each was to have its own garden, divided from its neighbours by hedges and which would contain a mix of housing types in which all classes could live. The plan was developed by the architect Raymond Unwin, who had worked closely with Ebenezer Howard. The housing stock of the resulting Hampstead Garden Suburb was heavily influenced by the Arts and Crafts movement, a reaction against the influence of the industrial revolution on urban life which called for simpler designs based on the traditions of the British countryside.

Against stiff opposition, Henrietta Barnett secured the purchase of the necessary land by 1907, and the construction of Hampstead Garden Suburb began the same year. In the century since then the Hampstead Garden Suburb Trust has resisted attempts to modify the essential character of the suburb and held back the encroachments of would-be developers. As a result, although increasing property prices have meant that it is perhaps less socially diverse than Henrietta Barnett originally intended, the suburb stands as a testament to her and to other Victorian social housing reformers who were determined that decent housing should not be the sole preserve of the affluent.

see more on next page >

WOOD

WOOD

MVTTON BROOK

FINCHLEY ROAD

BRIDGE LANE

TEMPLE FORTUNE LANE

FINCHLEY ROAD

HOOP LANE

~HAMPSTEAD · GARDEN · SVBVRB ·
~LONDON · N·W·

~BARRY · PARKER AND ~RAYMOND · VNWIN.
ARCHITECTS~
~HAMPSTEAD AND ~LETCHWORTH.
IN CONSVLTATION WITH
~Mᴿ E·L·LVTYENS.

~DRAWING Nᵒ 7809 APRIL 1911.

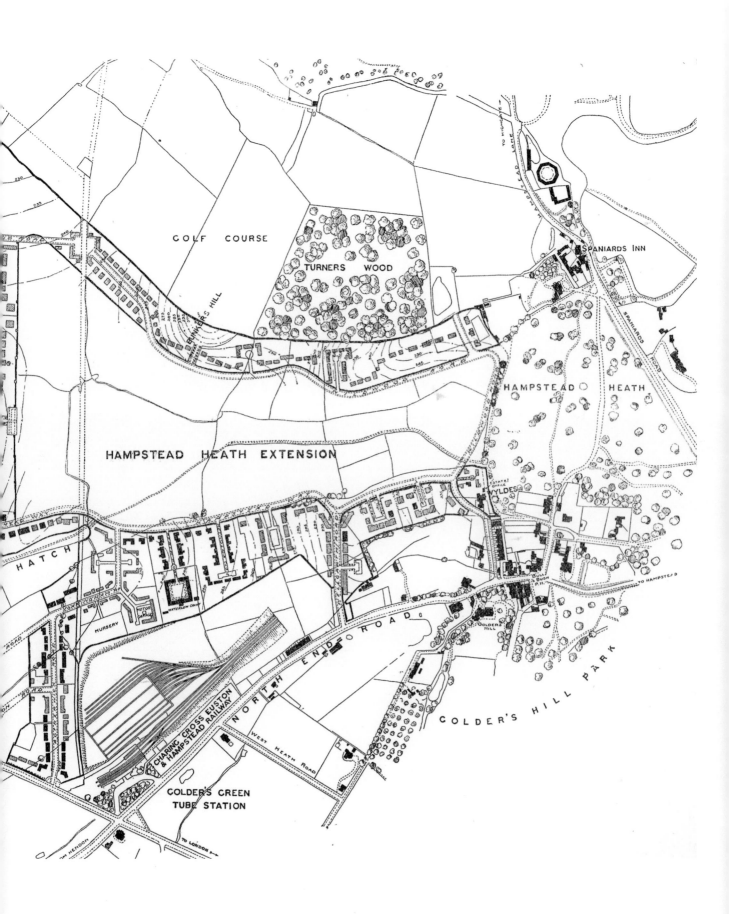

GOLF COURSE

TURNERS WOOD

BUNKER'S HILL

SPANIARDS INN

SPANIARDS

HAMPSTEAD HEATH

HAMPSTEAD HEATH EXTENSION

WYLDES

ESTATE OFFICE

HATCH

BULL & BUSH P.H.

TO HAMPSTEAD

GOLDERS HILL

MOTOR HOUSE

WATERLOW COURT

NURSERY

NORTH END ROAD

COLDER'S HILL PARK

CHARING CROSS EUSTON & HAMPSTEAD RAILWAY

WEST HEATH ROAD

GOLDER'S GREEN TUBE STATION

FROM HENDON

TO LONDON

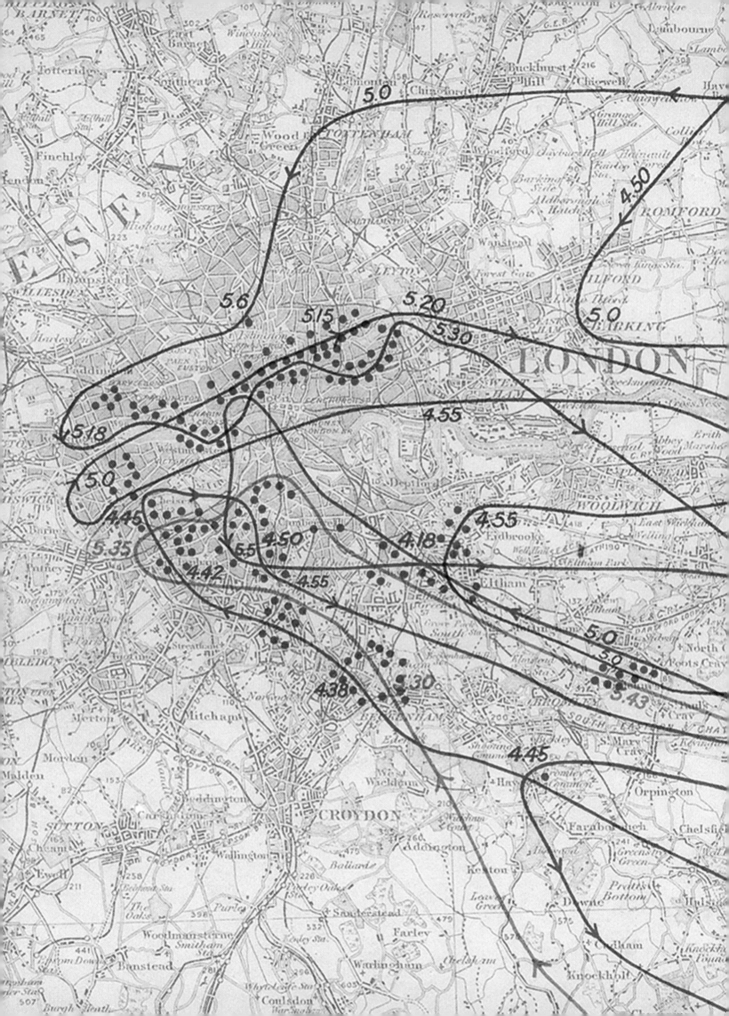

GERMAN GOTHA RAID MAP

The web of lines wavering across Kent and finally converging on London charted a new threat to Britain's security. They show the routes taken by German Gotha bombers as they launched an air raid on the capital on 6 December 1917. Britain, for so long reliant on its mastery of the seas to repel invaders, now faced attack from the air.

The assault had in fact begun nearly three years earlier, using Zeppelins, airships with envelopes filled with lighter-than-air but inflammable hydrogen gas. Developed initially by the German navy, the first Zeppelin raid had been launched against Great Yarmouth and King's Lynn in January 1915. The small bombs that were dropped killed four people, but caused panic as there seemed no effective defence against this death from the air. The Germans struggled at first to launch attacks on London as the Zeppelins' navigation was unreliable and one raid on London ended up in Hull. Even so, on 31 May 1915 London was hit for the first time by 120 bombs which hit in a strip between Stoke Newington and Leytonstone.

Mounting Zeppelin losses in 1916, as proper British anti-aircraft defences appeared (and British fighter pilots developed strategies to shoot down the airships), caused the German High Command to look for alternatives. The answer came in the form of the Gotha G.IV aircraft, which was more reliable, more manoeuvrable and could carry a greater bomb load than the Zeppelins. Operation Türkenkreuz ('Turk's Cross') was launched on 25 May, when twenty-three Gothas were sent to London. Heavy cloud caused them to divert and strike the area around Folkestone, where ninety-five people were killed. A second raid was similarly diverted to Sheerness, but a third, on 13 June, did reach London, where its bombs killed 162 people, including eighteen children who died when a bomb crashed through their classrooms in a Poplar primary school.

There was outrage and accusations that the Germans were resorting to using cowardly measures. But until effective counter-measures were developed, little could be done.

The first Gotha was shot down during a raid on 7 July. After that, anti-aircraft barrages and the improved skills of British pilots caused the Germans to resort to night bombing from 6 September.

The raids, though, continued and by the end of the war the Gothas had carried out fifty-two attacks and dropped 111 tons of bombs, killing over 800 people. The attack documented in the map was one of the less successful. Nicknamed the 'cock-crow raid' because of the unusually early timing of the raid (in the hours just after midnight on 6 December 1917), 5,000 pounds of bombs were dropped, causing £100,000 of damage and eighteen deaths, but good preparation by the Fire Brigade meant that the resulting fires were easily contained. Two Gothas were forced down during the raids by anti-aircraft fire: one German crew downed near Canterbury surrendered to the local vicar; another was allowed to land unhindered at an airfield near Rochford after giving, by chance, the correct identification signal. Four more of the Gothas were either shot down or crashed, making German losses of six out of sixteen.

The last raid took place on 19 May 1918, when thirty-eight Gothas and three of the even larger 'R-type' Giants – real behemoths with a wingspan of over 45 m (150 ft) – buzzed over London and killed forty-nine Londoners in a final night of agony. The Sopwith Camels of the London air defence squadrons and the anti-aircraft crews shot down six Gothas. With losses becoming unacceptably high, for the rest of the war the German airforce kept the Gothas back from further raids, while the centralization of Britain's previously separate air units into the Royal Air Force in 1 April 1918 further increased the British capacity to defend their airspace. There would be no more 'cock-crow' raids for the next twenty-two years until Britain's cities once more faced a devastating campaign of strategic bombing during the Blitz of 1940.

see more on next page >

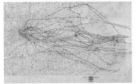

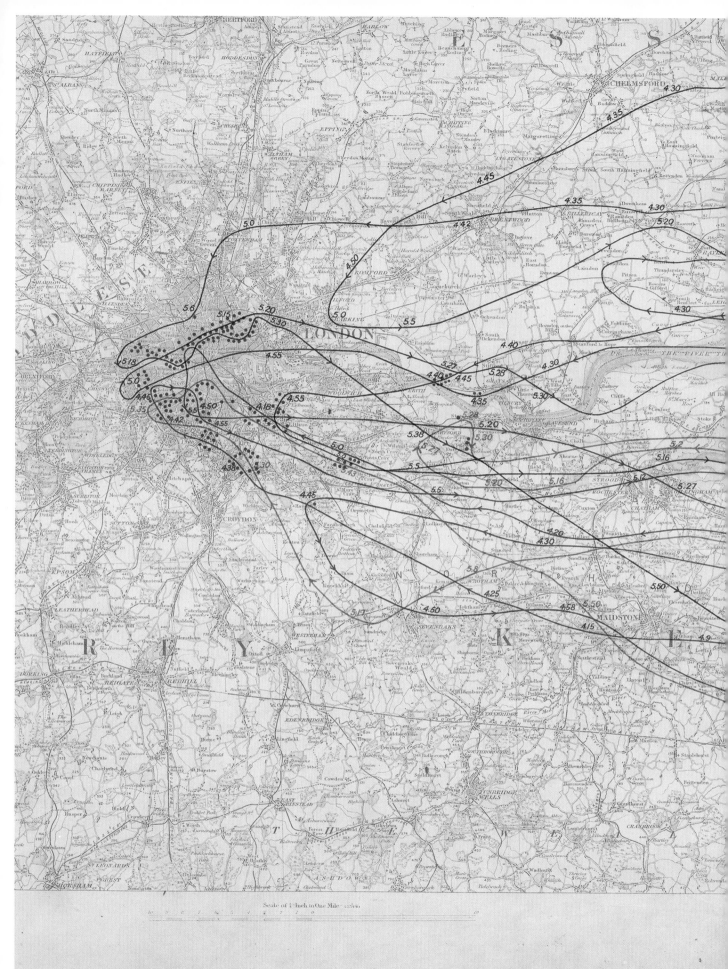

Scale of ¼ Inch to One Mile 1:253440

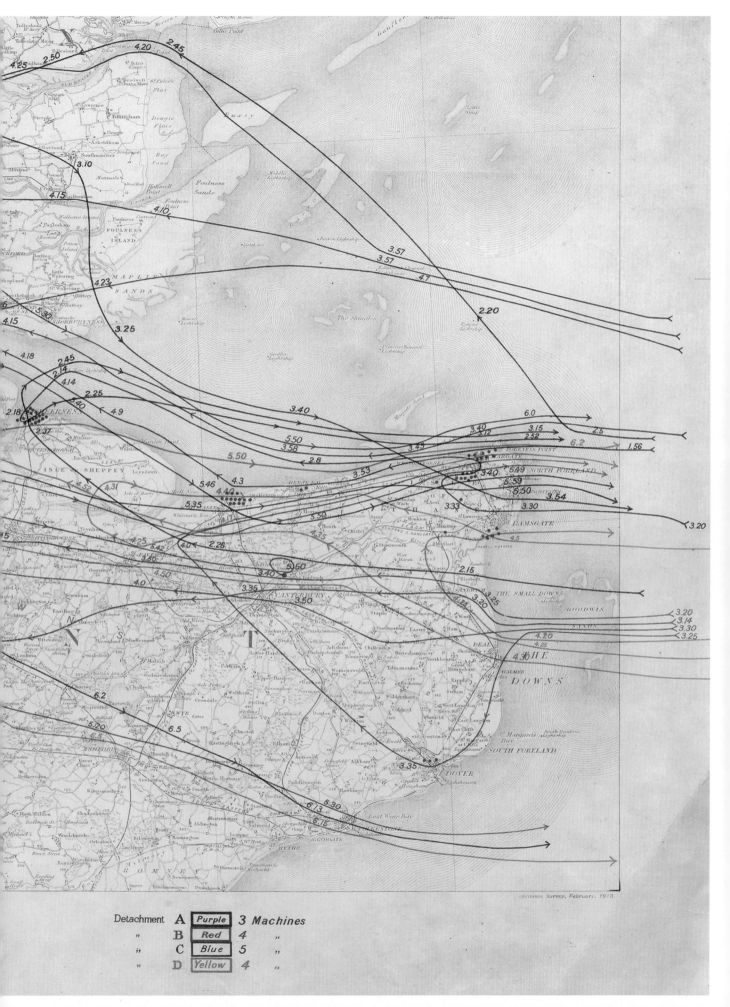

Detachment A **Purple** 3 Machines

" B **Red** 4 "

" C **Blue** 5 "

" D **Yellow** 4 "

GENERAL STRIKE PLAN

In May 1926, Britain experienced the only General Strike in its history, a titanic struggle between trade unionists and the government for which only one side was truly well prepared. The map formed part of the official planning in the event of such a strike, marking out key transport and power installations which needed to be defended and kept operating in order to defeat the strikers.

Since the origins of trade unions in the early days of the Industrial Revolution, British unions had made enormous progress, both in terms of the numbers of their members (which reached around 6.5 million by the end of the First World War) and the level of cohesion of the movement; the Trades Union Congress (TUC), established in 1868, had gradually achieved the affiliation of almost all the principal unions and acted as a powerful lobby for the rights of union members generally. The Labour Party, founded in 1900, provided political muscle for the unions – although they did not always agree on the same priorities for the labour movement – and in 1924 formed their first government (albeit a minority one), with Ramsay MacDonald as prime minister.

The fall of the MacDonald government after just nine months – swept away by a Conservative landslide – proved a disaster for the union movement. It came at a time of tension within the mining industry, where output was falling (down from around 300 tons annually per man in the 1880s to around 200 tons by 1924) and the mine-owners were determined to make cuts. In June 1925 they announced they would reduce rates of pay, abolish the minimum wage and reduce hours. While the General Council of the TUC announced it would support the miners, the new conservative government staved off a damaging dispute by guaranteeing it would provide the cash to make up for the miners' lost wages.

A Royal Commission was set up under Sir Herbert Samuel to examine the industry, but its report recommended removing this subsidy, leaving the miners' leaders with almost nothing. The mine-owners, emboldened, announced new terms which involved a pay cut of between 10 and 25 per cent. When negotiations between the government and the TUC broke down, a confrontation seemed inevitable. On 1 May, the TUC Conference announced a general strike, while the miners' union called for action under the slogan 'Not a penny off the wage, not a minute on the day'.

From the start the miners' position was weak. Mining production had increased in 1925 and power stations had enough stockpiles to keep electricity supplies running for eight to ten weeks. The government had also made plans, using its powers under the 1920 Emergency Powers Act to establish ten national areas, each with a commissioner and a junior government minister, to keep transport moving and food supplies being delivered. A militia, the Organisation for the Maintenance of Supplies (OMS) was established, with 100,000 volunteers to maintain law and order and keep the economy moving in the event of a strike. The map formed part of this planning – which included the sectioning off of a large part of Hyde Park with barbed wire to turn it into an enormous food depot. Barracks, territorial army bases and fire stations are marked – all the infrastructure which would be necessary to keep the capital moving in the event of a general industrial action by union members.

When the strike began on 4 May, the TUC was hopeful. Almost 1.75 million key workers walked out and at Paddington Station only three men turned up for work (and only a solitary train ran to Oxford). But the efforts of the OMS, white-collar volunteers and a handful of strike-breakers began to tell. The OMS recruited 644 train drivers, goods could still be carried by road, and a government-printed newspaper, the *British Gazette*, edited by Winston Churchill, promoted a staunch anti-strike message (while the government requisitioned all the paper needed by its union counterpart, the *British Worker*, cutting off the supply of news favourable to the strikers).

By the third day of the strike the tide had begun to turn. Buses appeared on London's streets and food supplies became more plentiful. On 12 May, as a stream of workers returned, the TUC went to Downing Street to negotiate a settlement over the heads of the miners' leaders. Prime minister Stanley Baldwin offered them no guarantees over wages or hours and the cancelling of the General Strike the same day was made to seem like an ignominious retreat.

The miners struggled on until November, by which time most of them had returned to work, having achieved none of their objectives. A damaging split in the union between moderates and militants then crippled the mining union for a decade. More generally, rising unemployment tempered the appetite for labour militancy (which only returned when the Second World War brought back full employment) and General Strikes were banned in 1927. But it had been the preparations that the British government had made to see off such an eventuality which had played a large role in the defeat of the 1926 General Strike, and the mapping which survives from their emergency planning is eloquent testament to the determination government ministers showed not to be defeated by the miners.

SOLAR ECLIPSE MAP

This 1927 map produced by the Ordnance Survey charts the predicted track of a total solar eclipse which crossed parts of the United Kingdom on 29 June that year. It was the first such event to be visible in mainland Britain for over 200 years, creating huge excitement about a phenomenon that had in previous centuries been viewed as a cause for dread.

Despite the anticipation, the eclipse was something of a disappointment. The crowds who gathered had less than thirty seconds to view the totality, and heavy cloud and high winds meant that the darkness they experienced was largely of a meteorological rather than an astronomical nature.

By 1927 it was well known that such eclipses are caused by the passing of the Moon directly between the Earth and the Sun (an event made very irregular by the tilting of the Moon's orbit in respect of that of the Sun). However, in past ages, there had been no such understanding, and an eclipse, the dying of the light, had been regarded as an extremely evil omen. One, on 16 March 1485 was associated with the death of Anne Neville, the wife of Richard III. Nearly 250 centuries earlier, the *Anglo-Saxon Chronicle* related an eclipse with horror when 'the Sun became as it were a three-night old Moon, and the stars about it at midday'. The chronicler believed that the event, which occurred in 1133 and came to be known as King Henry's eclipse had foretold the death of the monarch.

Henry I in fact died two years later, after eating an excessive quantity of lampreys while campaigning to put down a baronial revolt in Normandy. Although a generally effective ruler, who strengthened the royal justice system and began the establishment of a proper exchequer, he had failed to provide for a proper succession (his sole son, William, having perished when the ship on which he was returning to England from France sank outside Barfleur harbour).

A power struggle broke out between his daughter Matilda (who had retained the title Empress after the death of her first husband Emperor Henry V in 1125) and Henry I's nephew, Stephen of Blois (the son of William the Conqueror's sister Adela). Stephen, being present in England at his uncle's death used the advantage of incumbency to press his claim, but four years later his cousin the Empress crossed over to England and began gathering support among sympathetic barons.

Here, a solar eclipse intervened again – or so the superstitious would have it. The obscuring of the Sun recorded on 20 March 1140 by the chronicler William of Malmesbury was interpreted as meaning that Stephen would soon be ousted from his throne. And so it turned out (though a hefty dose of hindsight was probably involved in blaming the eclipse). At the Battle of Lincoln in February 1141, Stephen was defeated and captured. His cause would have been lost had not Matilda's half-brother Robert of Gloucester been taken at Winchester seven months later, and an exchange of noble prisoners resulted in their both being freed.

The war rumbled on for another thirteen years, with the Empress's partisans occupying much of the southwest of England and Stephen's supporters retaining control of the southeast. The debilitating stalemate was only brought to an end in 1152 when Matilda's son Henry returned to England with a fresh army. War-weary, Stephen agreed to accept him as his heir, as long as Henry in turn acknowledged him as his liege lord. There was no eclipse to blame when Stephen fell ill with a stomach disorder and, bleeding internally, died at Dover the following October. Virtually unopposed, Henry II ascended to the throne, bringing peace to England after eighteen years of uncertainty and civil war.

The viewers of the eclipse in 1927 would have known they were privileged, being the first to view a total solar eclipse in Britain since 3 May 1715. On that previous occasion the event had been documented by Edmond Halley, a renowned astronomical scholar, who would become the second Astronomer-Royal five years later. The understanding of the science behind planetary and satellite orbits (and thus of eclipses) had advanced hugely since the twelfth century – in part thanks to Halley's close friend Isaac Newton – and Halley published a map before the event predicting the course of the eclipse. He was only around 20 miles out in the boundaries he drew for its track and the crowds that gathered in London at 9 a.m. on a sunny spring morning were treated to a spectacular show. Georgian England, indeed, enjoyed an encore just nine years later, when a second total eclipse was visible in the south, accompanied by a new map by Halley (and a modified version showing the corrected track of the 1715 eclipse).

Halley was something of a virtuoso at predicting astronomical events. In 1705 he had published a work predicting the return in 1758 of a comet which had appeared in the skies around every 75 years since antiquity. It, too, had a baleful reputation, and its occurrence in the early months of 1066 were taken to presage a disaster for England (which duly arrived that October in the form of William the Conqueror's victory over King Harold at Hastings). Although he did not live to see his prediction vindicated (having died in 1742), Halley achieved posthumous immortality by having the comet named after him. Those who gathered to see the 1927 eclipse and whom the clouds denied the full drama of the sun's demise, would have known that few of them would be alive for the next show, which took place only in 1999.

ORDNANCE SURVEY MAP
OF THE
SOLAR ECLIPSE
29th June, 1927.

Crown Copyright Reserved

This map has been prepared in co-operation with the Joint Permanent Eclipse Committee of the Royal Society and the Royal Astronomical Society to show the path of the shadow of the Moon during the total eclipse, the magnitude of the partial eclipse visible outside the limits of totality, the Greenwich Mean Time of middle of eclipse, and the altitude of the Sun at that time.

Summer Time is one hour later

The curves have been calculated for the Committee by Dr L. J. Comrie, of the Nautical Almanac Office.

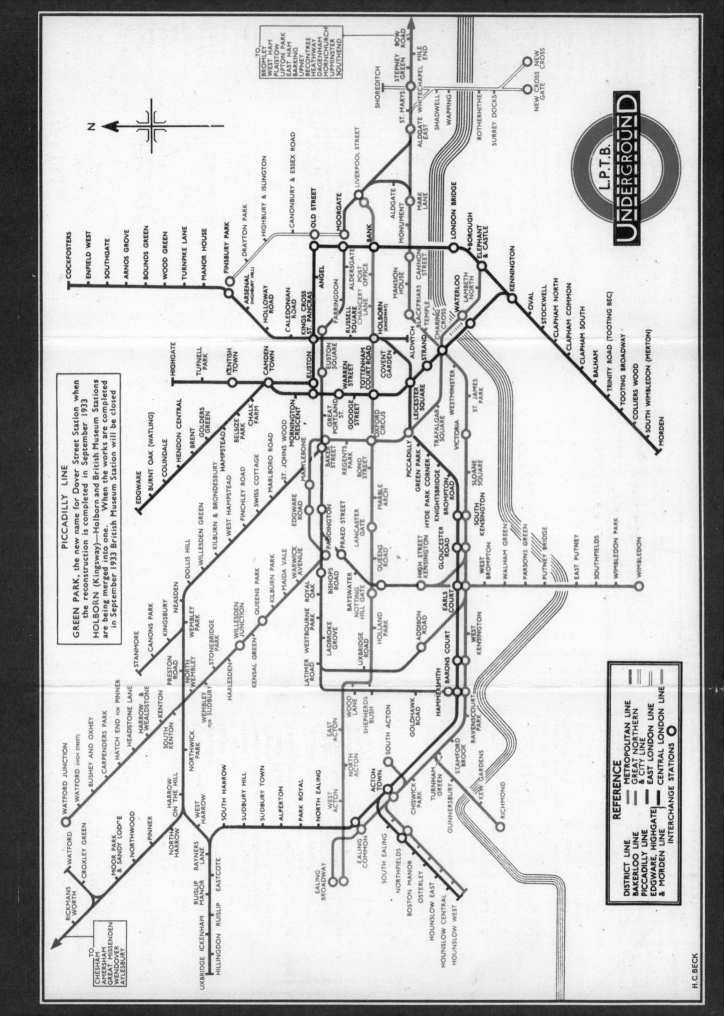

LONDON UNDERGROUND MAP, Harry Beck

Harry Beck's 1933 map of the London Underground is a design icon. Bringing a refreshing geometrical symmetry to what had been the confusing spaghetti-tangle of previous plans of the system, it has been the basis ever since for route diagrams of one of the world's great public transport networks.

Beck was a twenty-nine-year old junior draughtsman in the Signal Engineer's office when he submitted his plan for the map. The Underground had become a victim of its own success and its huge expansion since the opening of the first line in 1863. It was simply no longer possible to portray all the new lines which had opened without distorting them, bending them, having them change direction or even truncating the further-flung stops at end of each line. Or so everyone considered until Beck came along.

His solution was elegant, if a tad too radical for the Underground's publicity department which rejected his first draft in 1931. Taking the Central Line as a horizontal base, Beck used simple horizontals, verticals and diagonals to show the relationship between stops without any sense of the distance above ground between stops. That was irrelevant, he reasoned. Passengers needed to find their way around the system, not keep a track of how far they were travelling.

Two years later, the publicity department relented and 750,000 copies of Beck's map were printed. He was paid a mere ten guineas for his original artwork and a further five for that of a larger poster. However, he continued to work on almost all revisions of the map until 1960 (showing extensions such as that of the Piccadilly Line to Cockfosters in 1933, as shown in the example here, and the Northern Line to High Barnet in 1940). All subsequent iterations of the map have used Beck's design as their core. When travellers think of the Underground, they more often than not think of Beck's map.

What today's travellers may not think of are the origins of the Underground. Its opening in 1863 was the solution to a pressing problem for Victorian London: the population was simply growing too fast for its transportation system to cope with. Horse-drawn omnibuses took some of the slack, transporting 44,000 passengers daily (and, together with private horses contributing to the unfortunate need to dispose of millions of tons of horse manure each year). Most people simply walked – some 200,000 over Blackfriars and London Bridge each day.

Plans for an underground railway, the brainchild of Charles Pearson, a city solicitor, were approved in 1854 and his Metropolitan Railway received final parliamentary permission to open a line in 1859. Many thought the plan was absurd and that people would simply be asphyxiated in the cold and dark underground. However, work began in 1860 and thousands of navvies worked through the night to the light of flares to cut the tunnel from Bishop's Road Paddington to Farringdon. Finally, on 3 January 1863 the opening ceremony was held as 600 grandees took the first train (a leisurely affair that took two hours to complete the short route). Pearson, sadly, was not there to see his dream become reality, having died the previous year.

Thirty thousand people travelled on the new underground railway on its first day of public operation the following Saturday, paying between threepence and sixpence for a single ride in one of the three classes. Despite the choking fumes emitted by the steam trains which hauled the carriages, the Underground caught on and it carried 11.8 million passengers in its first year. The introduction of workmen's fares at threepence for a return, in 1864, enhanced its popularity, while the completion of the Circle Line in 1884 allowed it to become a true method of navigation around London, rather than a series of disconnected fragments of line.

There have been many innovations since. In 1890, electric trains were introduced, and the network expanded gradually, opening up new suburbs from which commuters could flood into central London, areas that were eulogized in London Underground posters from 1915 as 'Metro-Land'. In 1911 the problem of moving the growing numbers of passengers within stations was in part resolved by the opening of the first escalator at Earl's Court in 1911; reportedly the Underground management instructed 'Bumper' Harris, an employee who had a wooden leg following an accident, to travel up and down it all day to prove its safety. Passenger numbers rose inexorably, with a few bumps caused by rising car ownership and the Second World War, reaching 488 million in 1938 and 1.34 billion by 2016. For all of those passengers, the Beck map and its descendants proved an invaluable and stylish way of navigating the system.

BRITISH NATURAL RESOURCES

The map lays out Britain's agricultural, mineral and industrial resources, giving a colourful impression of a prosperous and self-reliant nation. Yet it conceals a difficult truth, for at the time of its production in the early 1940s, Britain was engaged in a life-and-death struggle to guarantee the importation of the strategic resources to enable it to continue the war against Germany.

Just as Napoleon before him, Hitler hoped to throttle Britain into submission by choking off its shipping lanes and preventing food and other goods from getting through. While Britain had adequate supplies of coal – it was one of the country's principal industries and production was at a more than adequate 4.3 million tons in 1940 – only around a third of food was produced domestically, and vital raw materials such as oil, cotton and iron all had to be shipped in by sea.

The instruments of Hitler's planned blockade were the U-boats of Admiral Dönitz's Kriegsmarine (the German navy). The very first U-boat attack on British shipping came on 3 September 1939 against a patrol boat, but the danger they posed grew exponentially after the fall of France in May 1940 brought her Atlantic ports into German control and allowed their use as U-boat bases. The period from June 1940 to February 1941 was referred to as the Kriegsmarine's 'Happy Time' as the introduction of 'wolf-pack' tactics, whereby the U-boats waited in groups and then swarmed onto convoys that passed their line, inflicting huge losses on British shipping. One convoy, code-named HX-79, in October 1940 lost twelve ships despite a strong escort of destroyers and corvettes.

Gradually, though, the British and then the Americans, after they joined the war against Germany in December 1941, turned the tide. The cracking of the German naval Enigma cipher after the capture of a machine on the crippled U-22 in February 1940 allowed British codebreakers to read the naval code throughout the summer and autumn of 1941, plot the course of the wolf-packs and take precautionary measures. The introduction of devices such as High Frequency Direction Finding radars, which allowed the interception of U-boat transmissions and the triangulation of their position, and Asdic, a form of sonar, which permitted their detection underwater helped protect the vital convoys. Finally, the basing of very long-range planes in Newfoundland in May 1943 closed the 'Mid-Atlantic gap' in which air cover for convoys had not been possible, and so left the U-boats with nowhere to hide.

The loss of thirty-four U-boats in May 1943 marked a turning point, and Dönitz called off the U-boat offensive in the Atlantic. A brief resumption in September 1943 proved disastrous, as a further thirty-four U-boats were sunk in four months, having hit fewer than half that number of Allied vessels. All in all, the Battle of the Atlantic had cost the Allies around 3,500 merchant ships and the loss of about 83,000 Allied sailors (navy and civilian). The losses were severe, but they were not catastrophic and Britain and the United States were able to replace shipping at roughly twice the rate of the losses they were incurring. Only around one in ten convoys were attacked, and so at least 90 per cent of supplies got through to the United Kingdom.

The British Information Service, which distributed this map, was established at the consulate in New York to steer Britain's propaganda effort as the Second World War began and to promote a favourable view of Britain. It played an important part in the effort to bring the United States into the war on the Allied side, a successful campaign which meant that the British could continue to import strategic goods and not rely on the rather rosy picture of its natural resources which the map portrays.

see more on next page >

GREAT BRITAIN

HER NATURAL & INDUSTRIAL RESOURCES

NORTH SEA

ATLANTIC OCEAN

JOHN O'GROATS

WICK

STORNOWAY

PORTREE

KYLE of LOCHALSH

MALLAIG

FORT AUGUSTUS

FORT WILLIAM

OBAN

INVERNESS

ELGIN

FRASERBURGH

PETERHEAD

ABERDEEN

MONTROSE

ARBROATH

DUNDEE

ST. ANDREWS

PERTH

EDINBURGH

GLASGOW

GALASHIELS

BERWICK

AYR

GIRVAN

STRANRAER

DUMFRIES

CARLISLE

NEWCASTLE

DURHAM

DARLINGTON

MIDDLESBROUGH

ISLE of MAN

BELFAST

ANTRIM

LONDONDERRY

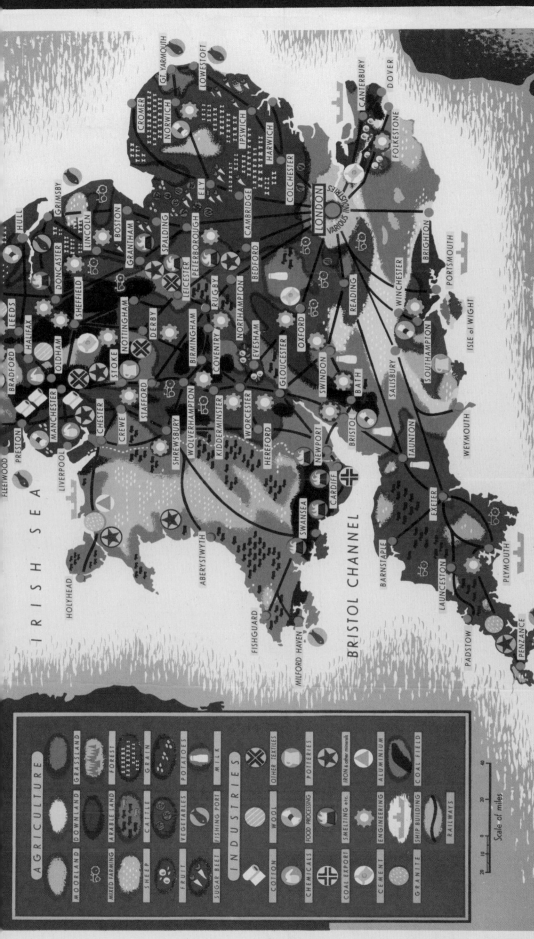

ENGLISH CHANNEL

IRISH SEA

BRISTOL CHANNEL

ISLES OF SCILLY

LAND'S END

AGRICULTURE

GRASSLAND
FOREST
GRAIN
POTATOES
MILK
MOORLAND
DOWNLAND
ARABLE LAND
CATTLE
VEGETABLES
FISHING PORT
MIXED FARMING
SHEEP
FRUIT
SUGAR BEET

INDUSTRIES

OTHER TEXTILES
POTTERIES
IRON & other minerals
ALUMINIUM
COAL FIELD
WOOL
FOOD PROCESSING
SMELTING etc.
ENGINEERING
SHIP BUILDING
RAILWAYS
COTTON
CHEMICALS
COAL EXPORT
CEMENT
GRANITE

Scale of miles

HOLYHEAD
FLEETWOOD
PRESTON
LIVERPOOL
FISHGUARD
MILFORD HAVEN
ABERYSTWYTH
SWANSEA
CARDIFF
NEWPORT
BARNSTAPLE
PADSTOW
PENZANCE
LAUNCESTON
PLYMOUTH
EXETER
WEYMOUTH
TAUNTON
BRISTOL
BATH
SALISBURY
SOUTHAMPTON
ISLE of WIGHT
PORTSMOUTH
BRIGHTON
WINCHESTER
READING
SWINDON
GLOUCESTER
OXFORD
EVESHAM
HEREFORD
WORCESTER
KIDDERMINSTER
WOLVERHAMPTON
SHREWSBURY
STAFFORD
CREWE
CHESTER
MANCHESTER
OLDHAM
LIVERPOOL
HALIFAX
BRADFORD
LEEDS
HULL
GRIMSBY
DONCASTER
SHEFFIELD
LINCOLN
BOSTON
GRANTHAM
NOTTINGHAM
DERBY
STOKE
BIRMINGHAM
COVENTRY
LEICESTER
RUGBY
NORTHAMPTON
BEDFORD
CAMBRIDGE
PETERBOROUGH
SPALDING
ELY
COLCHESTER
IPSWICH
HARWICH
NORWICH
CROMER
GT. YARMOUTH
LOWESTOFT
LONDON
VARIOUS INDUSTRIES
CANTERBURY
DOVER
FOLKESTONE

Distributed by the British Information Services, an Agency of the British Government, 30 Rockefeller Plaza, New York City

OPERATION SEA LION PLAN

It is a map of an invasion that never came. This German military map from the summer of 1940 marks the landing sites for the three successive waves of troops that were to capture a broad swathe of the Sussex and Kent coasts before pushing inland to threaten London. Fortunately for the British, overambition and the failure by the German Luftwaffe to establish air superiority meant that the German plan, Operation Sea Lion (or *Seelöwe*), was never carried out.

Britain's declaration of war on Germany in September 1939 after the Nazi invasion of Poland had always rankled with Adolf Hitler, who had considered the British to be potential allies. Although he hoped that a succession of defeats would lead the British to negotiate, they showed no signs of doing so, and when the defeat of France had brought Germany possession of the French Channel ports closest to southern England, in May 1940, an invasion was at last practicable.

Winston Churchill, who became British prime minister in May 1940, showed even less inclination to have truck with talk of surrender. On 16 July Hitler issued Führer Directive 16 declaring that: 'As England, in spite of her hopeless military situation, still shows no signs of willingness to come to terms, I have decided to prepare, and if necessary to carry out, a landing operation against her. The aim of this operation is to eliminate the English Motherland as a base from which the war against Germany can be continued.'

For this to be achieved, however, the Royal Air Force had to be defeated and control of the English Channel achieved by sweeping it of mines and keeping the Royal Navy fleets in the Mediterranean and North Sea from intervening. The last two proved easier to achieve, but, as Göring, the head of the Luftwaffe pointed out, the first might prove impossible. An invasion day was set for around 25 August 1940 and an aerial assault ordered on England in preparation. *The Adlerangriff* ('Eagle offensive') was launched with *Adlertag* ('Eagle Day') attacks on RAF stations throughout the southeast of England.

Yet neither this – in which the Luftwaffe lost about forty-seven aircraft to the RAF's twenty-four – nor subsequent attacks succeeded in destroying the RAF, which used a chain of recently constructed radar stations to gain warning of approaching Luftwaffe squadrons and so deploy its response more effectively. By the time the air assault was diverted to strategic bombing of cities such as London, beginning the Blitz, on 10 October, the Germans had lost over 1,600 aircraft and had never come close to achieving command of the air.

The loss of the this air war, the Battle of Britain, was a fatal blow to Sea Lion. It was difficult enough that the plans called for the landing of a first wave of 67,000 troops and a second wave of 71,000 inside two days and that the lack of specialist landing craft had meant the commandeering of around 2,400 river barges from throughout Nazi-occupied Europe. To transport the men of the Sixth, Ninth and Sixteenth Armies across the Channel under close air attack would invite a massacre.

The invasion had never had many supporters inside the German military hierarchy; on 13 August General Jodl, the army Chief of Operations Staff had written it was an assault that could only be launched out of desperation. The plans were gradually pared down from a broad front assault on the area between Ramsgate and the Isle of Wight, to a much tighter front focusing on the coast from Eastbourne to Folkestone, with an airborne attack intended to seize Dover, as shown in this map.

On 14 September, however, a planning meeting put back the invasion still further – it had been intended to take place the following day. Three days later Hitler ordered that the invasion fleet be dispersed as it was suffering unacceptable losses (and, indeed, on 18 September an RAF attack on Dunkirk destroyed twenty-six barges). By 12 October, Hitler had ordered that the troops earmarked for Operation Sea Lion be released for other fronts.

Sea Lion was now effectively dead, though Hitler only finally ordered a stop to planning on 23 September 1941. No longer would Britain be divided into six military-economic zones, as the political annex to Sea Lion had called for, based on a German military government at Blenheim Palace (a deliberate snub to Winston Churchill, whose ancestral home it was). Nor would the Gestapo swoop in and arrest any of the 2,820 people (politicians, trade unionists and army officers) deemed most dangerous to the occupation or the German army be able to install a Quisling regime headed by compliant fascists such as Oswald Mosley. That the plans had existed, though, was sobering enough and that they had failed was in large part due to the staunch resistance offered by a handful of RAF pilots.

see more on next page >

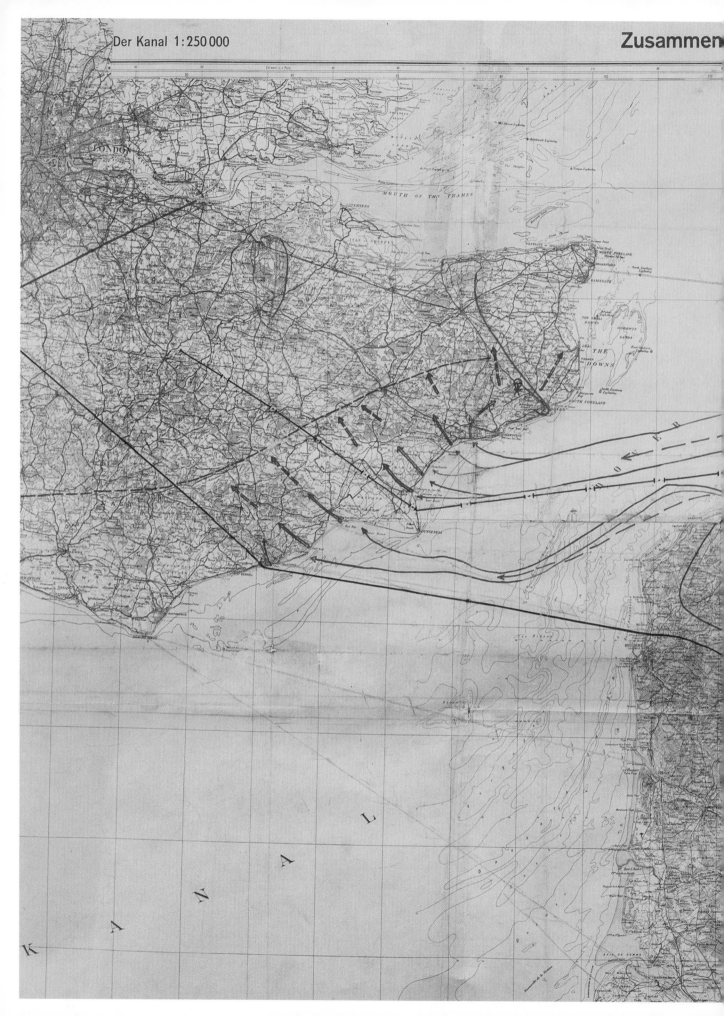

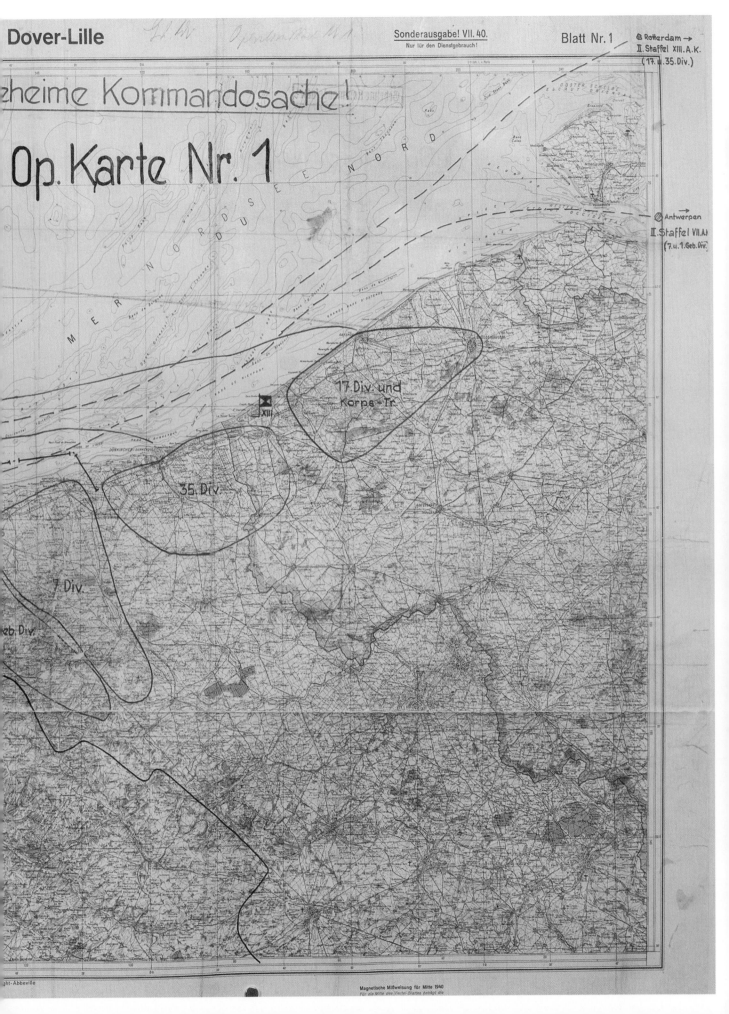

⊗ Rotterdam →
II. Staffel XIII. A.K.
(17. u. 35. Div.)

zheime Kommandosache!

Op. Karte Nr. 1

→ Antwerpen
II. Staffel VII.A.
(7. u. 1. Geb. Div.)

17. Div. und
Korps-Tr.

XIII

35. Div.

7. Div.

eb. Div.

GERMAN TARGET MAP OF LIVERPOOL

The yellow areas on this map mark the targets for the German bombers which attacked Liverpool during the Blitz of 1940–41. The city suffered a series of devastating raids, which made it the most heavily bombed conurbation in England outside London, and inflicted thousands of casualties on the inhabitants of Merseyside.

Once the Luftwaffe, the German air force, had been denied by the Battle of Britain the air superiority which would have allowed German forces to launch an invasion of southern England, its political master, Adolf Hitler, ordered its bombers to turn to a campaign of strategic destruction of British cities. By the time the Blitz was launched against London in early September 1940, other urban centres had already been struck. Liverpool, which possessed the largest port on the west coast, and was a vital conduit for food and other vital supplies being shipped across the Atlantic from Canada and the United States, was an obvious target.

The city's torment began on 28 August 1940 when 160 Luftwaffe bombers attacked the strip of dock installations along the Mersey waterfront. On the map these are marked with a yellow band, indicated by *Ziele* ('targets') in the key, while Luftwaffe pilots are warned away from the Internment Camp at Huyton ('*Internierungslager*') where 5,000 Austrian and German civilians had been held since the start of the war.

Most of the internees were released by May 1941, by which time the worst of the bombing was over. The damage, however, had been terrible. In the worst incident, on 29 November 1940, a German parachute bomb score a direct hit on a school in Durning Road, Edge Hill, where 300 people were sheltering in a basement bomb shelter. The building collapsed directly on top of them, while scalding water flooded the basement from the boiler situated there. Crushed and drowned, at least 166 people died, many of them children.

A three-day bombing campaign at Christmas 1940 destroyed the headquarters of the Cunard Line, while a renewal of the Blitz in May 1941 involved nearly 700 Luftwaffe bombers, which dropped 2,300 high-explosive bombs, put half of the docks out of action and destroyed 6,500 houses. Among the buildings damaged was the city's Anglican Cathedral, while one dock (Huskisson No. 2) was utterly devastated when the SS *Malakand*, a cargo ship carrying munitions, exploded after being hit by an incendiary bomb.

In all, the Liverpool Blitz killed nearly 4,000 people before Hitler's attack on the Soviet Union in June 1941 caused the definitive cancellation of plans to invade Britain and sucked resources into the eastern maelstrom. Winston Churchill paid tribute to the city's fortitude, commenting after one visit there following a raid: 'I see the damage done by the enemy attacks, but I also see . . . the spirit of an unconquered people.'

In 1932, the British prime minister Stanley Baldwin had declared 'the bomber will always get through'. It was a declaration that much of the next war was expected to be on the level of strategic bombing and that little could be done to prevent this. In fact, dozens of Luftwaffe planes were shot down by anti-aircraft fire or patrolling RAF fighters and the damage inflicted on Liverpool's shipping industry, while serious, was never fatal to the war effort. A prime requisite for successful strategic bombing was command of the air space over the target. The Battle of Britain had stopped Hitler achieving that, and so, while the yellow targets on the Luftwaffe map mark Liverpool's suffering, they did not foreshadow its destruction.

BLITZ BOMB DAMAGE MAP

London is a patchwork of coloured segments in this map produced by London County Council's architects' department to document the damage caused by the Blitz, the eight-month long aerial assault by Germany's Luftwaffe, which destroyed around a million houses and left large sections of the capital devastated.

'The bomber will always get through', Stanley Baldwin, the then British prime minister, had proclaimed in 1932, echoing a generally held view that in the event of another war, strategic bombing would play a huge role in the conflict and that preventing it from doing so would be impossible. At first, though, the attacks failed to materialize. A long period of deceptive calm (known as the 'Phoney War') was followed by the lightning German land assaults on Belgium, the Netherlands and France, but still German bombers largely kept away from British airspace.

It was only when an invasion of Britain became a practical proposition for the Germans after their occupation of northern France in May 1940 that attacks by the Luftwaffe began in earnest. Yet these were at first concentrated on destroying the RAF as a means of establishing the air superiority necessary to allow the transportation of German troops across the English Channel. It was only when, towards the end of August, that it became clear to Hitler that Germany was losing this Battle of Britain, that he ordered his fleet of bombers to turn their attention on British cities as a means of cracking the country's morale and forcing its surrender.

On the afternoon of 7 September 1940 a huge bomber force, some 600-strong, dropped wave after wave of bombs and incendiaries on East London, concentrating on the docks. Woolwich, Millwall, Beckton, Limehouse and Rotherhithe were all pounded in the first attack of the Blitz, a strategic bombing campaign that would last eight months and expand to include most of Britain's principal ports and industrial cities.

The attacks on London in particular were almost relentless, as the bombers returned night after night for nearly eight weeks. The capital's defences were nearly useless; anti-aircraft guns were only effective against low-level attacks and when the Germans switched to night bombing, the British night fighters struggled to counter them until the introduction of the Bristol Beaufighter late in the Blitz. Tens of thousands of Londoners huddled in deep shelters, including in Tube stations, as the bombs fell.

But still morale did not crack, despite the enormous losses. On 14 October, some 400 bombers struck, killing 200 people, and in that month alone around 9,000 tons of explosives rained down on London. On 29 December in a huge raid, a firestorm engulfed the fringes of the City of London, the flames consuming the area around St Paul's Cathedral, but never setting fire to the iconic building. A small army of Air Raid Precautions (ARP), Auxiliary Fire Service (AFS) and Women's Voluntary Services (WVS) officers kept order, doused the flames, ensured houses were blacked-out to avoid guiding the German bombers in and helped clear the rubble afterwards.

The raids then began to fall off, as the Germans, having cancelled plans to invade England in September 1940, switched resources to other fronts, including the Balkans and the planned offensive against the Soviet Union (which finally took place in June 1941). The last major raid on London on the night of 10–11 May 1941 was one of the worst, hitting a broad swathe of the city from the docks in the east to the West End. The casualties were the highest of any night's bombing: 1,436 killed and 1,800 wounded, with over 11,000 houses destroyed. Almost the entire rail transport network was halted and the tangled mess of fractured gas pipelines and broken electricity and telephone cables took weeks to repair.

Forty-one thousand lives had been lost, but the nightmare for London (and within a month the rest of the country) was over. At least, that is, until Germany launched the V-weapon offensive on southern England in 1944–5 which saw the world's first ballistic missiles strike the capital. In the meantime, county council surveyors carefully documented the damage, colouring their maps with black for total destruction, purple for damaged beyond repair, red for degrees of serious damage, orange and yellow for minor damage and green for undamaged areas that would still have to be cleared.

Street

PICCADILLY

ST JAMES'S ST.

RITZ
HOTEL

Green Park Stn.

ST. JAMES'S PALACE

CLARENCE HOUSE AND
YORK HOUSE

Green
Park

Constitution Hill

Buckingham Gate

BUCKINGHAM
PALACE

Bi

ons will make the
t 9.30 a.m. With

MAP OF THE CORONATION PROCESSION

The map commemorates a day which brought a sense of relief to Britain after the trials of the Second World War and the grinding years of austerity which had followed it. The Coronation of Elizabeth II on 2 June 1953 was a moment for the nation to celebrate with lavish ceremonial and age-hallowed pageantry. The procession through London's streets, which followed the coronation ceremony itself, snaked through the city's historic heart, beginning at Westminster Abbey, before arriving back at Buckingham Palace nearly two hours later.

The Second World War had swept away much that was familiar; not just the many buildings brought low by the German Blitz (which killed 40,000 civilians in London alone), or the nearly 400,000 soldiers who had lost their lives during the conflict, but the sense that an old world had passed. This sentiment contributed to the landslide win by Labour's Clement Attlee in the 1945 general election, and the defeat of the veteran Winston Churchill, who had led the nation to victory and whom most had assumed would preside over it in peace.

The Attlee government acted on the 1942 Beveridge Report and established the National Health Service, which began to operate from June 1948 and led to a revolution in the nation's healthcare. Some other things were slower to change. Much to the frustration of many, rationing continued to operate – and was in some cases intensified – so that bacon rationing only ended in July 1954, and clothing purchases still had to be conducted with clothes ration coupons, forcing the adoption of a sparse and austere style. In housing there were grave shortages, partly caused by bombing, but in large part by the lamentable state of the pre-war housing stock, while the reintegration into the economy of millions of service personnel who had been demobilized caused severe disruption.

There was a yearning for something new, for an escape, a need partly met by the revival of the Ideal Home Exhibition at Olympia in 1947 and the arrival of the 'New Look', a fashion trend that tried to make the most of limited means with long, swirling skirts. Conversely, there was a political flight to safety when the Labour government was defeated in the 1951 general election and the ageing, yet familiar, face of Winston Churchill returned to Downing Street.

That summer the nation had celebrated the Festival of Britain, intended to mark the centenary of the Great Exhibition of 1851.

A World's Fair without the rest of the world, it unashamedly showcased Britain, with displays of the latest advances in science, technology and design and the best of British arts and crafts on London's South Bank and at a variety of venues around the country. A slender, aluminium-clad, cigar-shaped structure, the Skylon, some 90 m (300 ft) high, presided over the Festival, which more than succeeded in its message that, as well as being an old-established and respected power, Britain was a youthful and vigorous contender, too.

George VI, who had been the nation's figurehead throughout the war, died on 6 February 1952, leaving his twenty-five-year old daughter Elizabeth to succeed him. The planning for the new queen's coronation began almost immediately, but the actual ceremony only took place 14 months later. British coronations were always occasions for pomp and display, but Elizabeth II's took place under the glare of unprecedented publicity. After a bitter behind-the-scenes argument, television cameras were allowed to film the ceremony, leading to the event being watched by over 20 million viewers (at a time when there were only 2.7 million television sets in Britain, many of them bought in order to be able to watch the coronation).

The three-million strong crowd who lined London's streets did not see the Queen enter Westminster Abbey preceded by St Edward's Crown, based on a medieval original, and used since the coronation of Charles II in 1661, nor her gown designed by Norman Hartnell embroidered with emblems from the main Commonwealth Countries (a Tudor rose for England, a maple leaf for Canada, a wattle for Australia and a lotus flower for India), but they did catch a glimpse of the fairy-tale carriage in which she was carried down Whitehall, Piccadilly, Oxford Street and Regent Street along the circuitous 8 km (5 mile) route back from the Abbey to the Palace. The line of the 10,000 servicemen who marched as part of the parade stretched for almost two miles.

It seemed truly the beginning of a new Elizabethan age, filled with hope. That news of Tenzing Norgay and Edmund Hillary's first ascent of Mount Everest reached London on Coronation Day only added to the excitement. That the war had almost bankrupted Britain, that her colonies were restive and that some, such as India, were already independent countries, and that, economically and diplomatically, the country was falling behind the United States and the resurgent nations, these were matters for the future.

see more on next page >

In addition to the Royal processions, the Lord Mayor of London will leave the Mansion House at 7.55 a.m. to drive in his State Coach drawn by six horses along the Embankment, joining the main Processional route at 8.30 a.m. at Hungerford Bridge, and arriving at the Abbey at 8.45 a.m. Led by the Marshal of the City of London, mounted, and followed by the Lord Mayor's footmen in their liveries, the Lord Mayor will be accompanied by the Lady Mayoress and the Common Cryer with Sword and Mace, and will have an escort of pikemen.

The Speaker of the House of Commons will make the traditional short drive to the Abbey at 9.30 a.m. With him in the Coach will be the Serjeant-at-Arms with the Mace, and the Speaker's Chaplain. There will be an escort of one Life Guardsman and, walking before, the Speaker's Secretary and Trainbearer.

B.B.C. Observers will be stationed at Buckingham Palace, Trafalgar Square, Westminster Abbey and the Annexe, Pall Mall, Marble Arch and Piccadilly Circus. There will be television cameras at Buckingham Palace,

PROCESSION
QUEEN ELIZABETH II

the Enbankment, Hyde Park and inside the Abbey. The troops lining the route will be: *The Mall:* Brigade of Guards; *Trafalgar Square to the Abbey and back:* Royal Navy, with Officer Cadets of all three Services in *Parliament Square; Cockspur Street:* Canadian Military Forces; *Pall Mall to Marble Arch:* the Army; and *Oxford Street back to the Haymarket:* Royal Air Force. Below are the approximate times of the return Procession.

	Depart Westminster Abbey	Trafalgar Square	Hyde Park Corner	Marble Arch	Oxford Circus	Piccadilly Circus	Arrive Buckingham Palace
Head of the Procession:	—	—	—	2.55	3.15	3.25	3.45
Her Majesty The Queen:	2.50	3.05	3.25	3.40	4.00	4.10	4.30

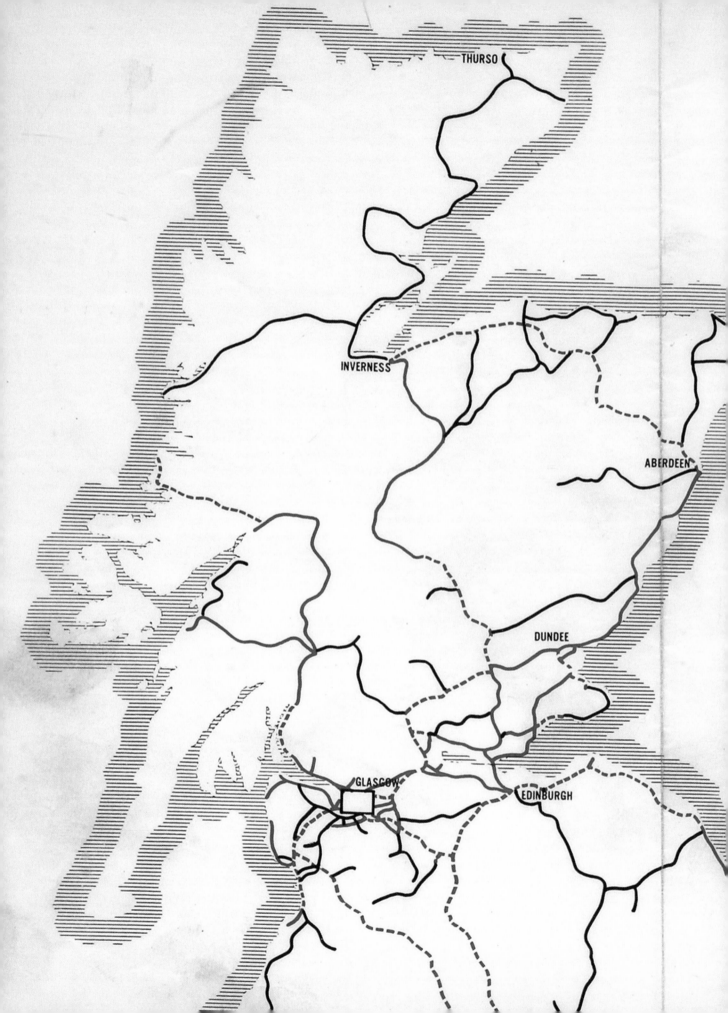

BEECHING CUTS MAP

The map unveiled by British Transport Commission Chairman Sir Richard Beeching in March 1963 marked the symbolic end of the great railway age. Taking an axe, as he did, to great swathes of rural lines, Beeching tried to fend off the challenge posed by the growth of road transportation and left large areas of the countryside with no train services at all.

The spider's web of track shown in his report, *The Reshaping of the British Railways*, with black indicating those routes which were to be closed, and red the lines which had been selected for survival (though not all with stopping services), was Beeching's way of solving a problem which had been apparent for some time: the railways were simply not profitable. Ticket revenues and freight charges were hopelessly inadequate to defray the expenses of running a comprehensive network, particularly as successive governments had taken the way of political least resistance by acceding to demands for higher wages in the rail industry, while at the same time keeping a ceiling on fare increases.

There had been some rationalization already – the rail network had reached its peak extent in 1914, when there were some 37,660 km (23,400 miles) of passenger track, and around 2,100 km (1,300 miles) had been closed by 1939. After the nationalization of the railways in 1948, the new British Transport Commission had pared down a further 5,300 km (3,300 miles) by 1962. But the salami-slicing of selected lines could not stem the losses, which had reached £104 million in the year before Beeching's report. Although there had been a fitful attempt at investing the railways out of trouble, when the 1955 Modernisation Report promised £1.24 billion of spending, this investment never fully materialized. To make matters worse, increasing car ownership and the flexibility that road transport offered, both to industries which could have freight carried directly to their doorsteps, and to individuals, who could travel whenever and wherever they wanted, was draining further business away from the trains. Many feared that the unspoken policy of combining unstoppable costs with an unmovably large network would lead to the death of the railways.

Beeching, on secondment from ICI, at the time Britain's largest manufacturer, stepped in with a solution that was as unpalatable as it was logical. *The Reshaping of the British Railways* provided a stark analysis: that 50 per cent of Britain's rail routes provided only 2 per cent of its revenues, and that half of the 4,300 stations had annual receipts of less than £10,000. Some lines covered barely 10 per cent of their running costs. To save the arterial rail routes, Beeching proposed ripping out the veins, ruthlessly shutting down railway tracks where there was no prospect of a profitable service, or where routes where duplicated. He earmarked 2,363 stations (just over half those operating in 1962) and some 8,000 km (5,000 miles) of track (about a third of the total) for closure.

There were howls of protest in the areas worst affected and a few lines in Scotland, Wales and the southwest of England were reprieved. Some stretches were saved and turned into heritage railways (or, later, into cycle paths) but most remained neglected and grassed over as monuments to what had once been the nation's most popular form of transport. The losses that Beeching had hoped to stem continued – it was estimated closing a third of the network had only saved £30 million a year – and were still running at £100 million in 1968. That year, a new Transport Act accepted that the railways would need to receive a subsidy for at least three years.

The British government never did rid itself of the need to subsidize the country's rail system, however. Even after the privatization of the network in 1994–7, the subventions continued, reaching £4.8 billion in 2014–15. Yet the railway network avoided complete collapse, and in terms of passenger numbers, prospered, so that passenger journeys (which had declined from 965 million a year in 1962 to 835 million in 1965 in the immediate aftermath of the Beeching cuts) reached a record 1.32 billion in 2010. Beeching's axe may have wounded the railways, but his blood-letting ultimately allowed them to survive.

see more on next page >

Map No.9

BRITISH RAILWAYS
PROPOSED WITHDRAWAL OF
PASSENGER TRAIN SERVICES

All passenger services
to be withdrawn

All stopping passenger
services to be withdrawn

THURSO

INVERNESS

ABERDEEN

DUNDEE

EDINBURGH

GLASGOW

NEWCASTLE

MIDDLESBROUGH

CARLISLE

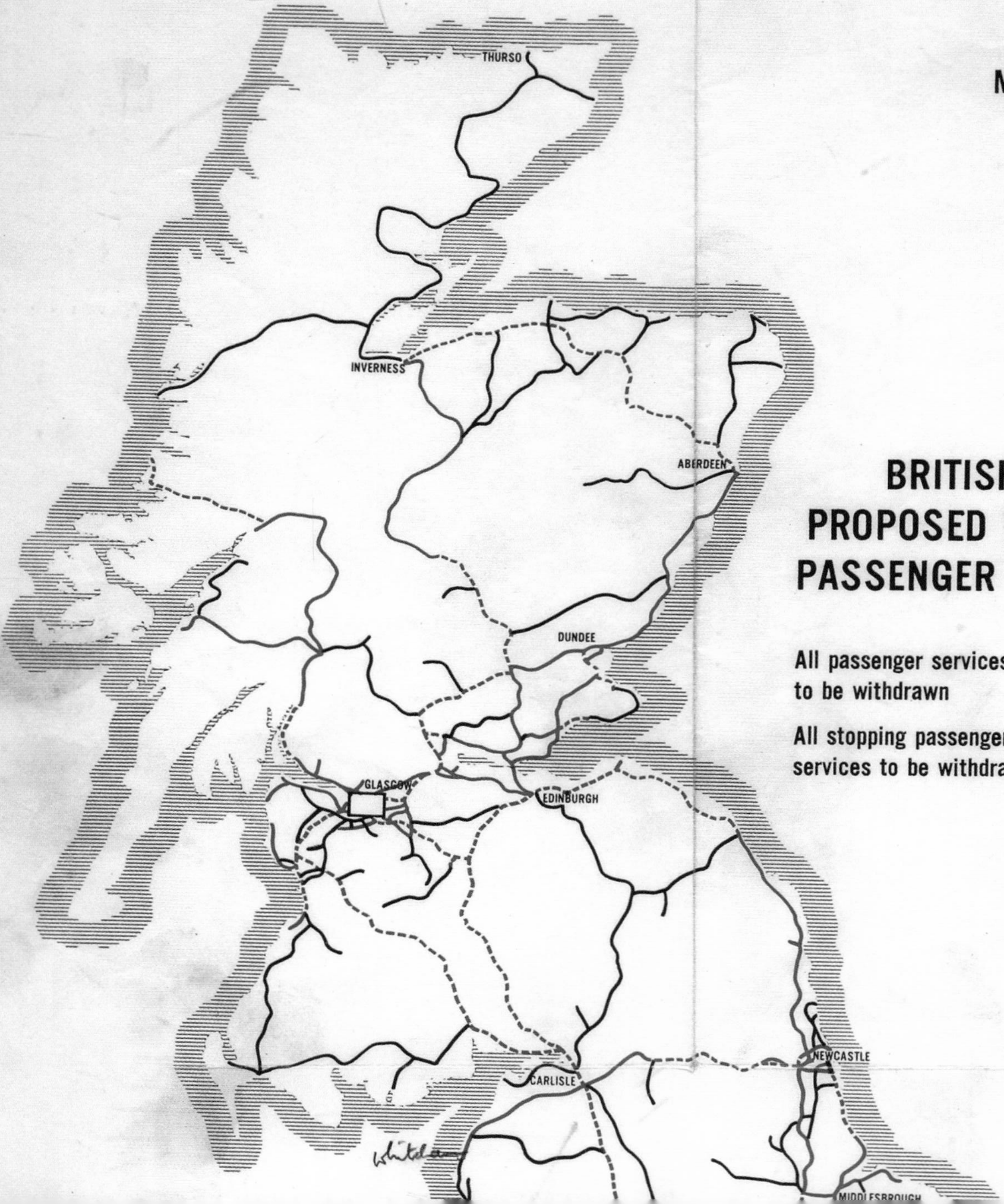

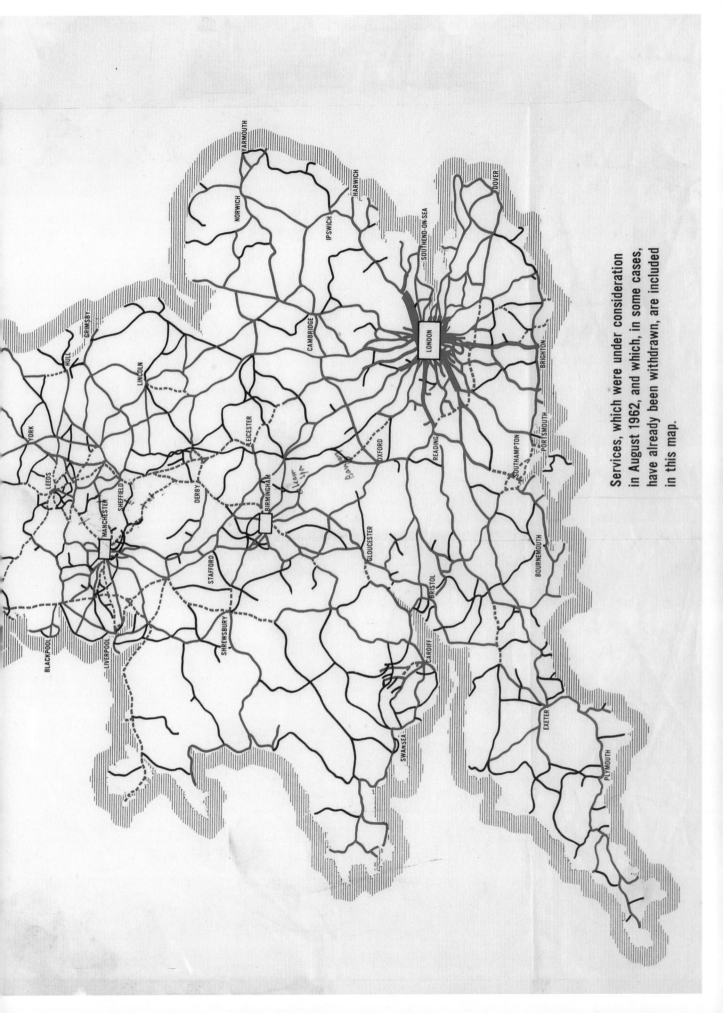

Services, which were under consideration in August 1962, and which, in some cases, have already been withdrawn, are included in this map.

AERIAL VIEW OF BATH

This extraordinary aerial photograph of Bath and its environs from 2001 has an almost map-like quality. The close-packed blocks of the town cling to the banks of the River Avon which spirals through it, as all around a swirl of fields and woods creates a green kaleidoscope of a precision that would have made the mapmakers of past centuries gasp.

Bath has a very ancient heritage. Medieval folklore had it that it was founded by Bladud, the son of Lud Hudibras, King of Britain (and a reputed descendant of the hero Aeneas, a refugee from the Trojan wars). The unfortunate Bladud contracted leprosy and was reduced to the humble status of an itinerant pig-herd. When the swine he was looking after contracted the disease, Bladud's ruin was complete and he fled, together with the leprous pigs. The herd, however, bathed in a muddy pool, and when the distraught prince washed the mud off, he found the pigs' leprosy was cured. Dipping in the waters himself, he found his own skin clear of the sores which had made him an object of fear and derision. The spring that had performed this miracle, so the story went, became a pilgrimage site for those seeking similar cures. Unfortunately for Bladud, he later became convinced that he could fly, had mechanical wings constructed and was dashed to pieces after he leaped from the top of a temple of Apollo.

What is certain is that by Roman times, there were flourishing baths, dedicated to the Romano-British goddess Sulis (after whom the town received its Latin name Aquae Sulis, 'the waters of Sulis') and devotees travelled from as far as north Britain, Gaul and Germany to bathe in the waters. Aquae Sulis was never a large town, but the steady stream of ancient health tourists who visited left their mark, including caches of 'curse tablets', invocations inscribed on lead including from spurned lovers and the simply vengeful. One calls down dark forces on a rival in love, praying that 'may he who carried off Vilbia from me become as liquid as water'.

After the fall of the Roman Empire, Bath went through hard times. It came into Anglo-Saxon hands around 577, after the native Britons lost control of the West Country and it took four centuries, until 973, when King Edgar was crowned there, for it to recover its position as an important centre. For the succeeding centuries, Bath Abbey dominated the town, but in the aftermath of its dissolution in 1539, the centre of the city decayed. The baths themselves were still operating, however. The antiquary John Leland, who visited in the 1540s, noted that they were 'much frequented of People diseased with Lepre, Pokkes, Scabes and great Aches', but noted, alarmingly, that they were never actually cleaned, but merely closed for three hours in the middle of the day, when it was assumed that the pools would, somehow, purify themselves.

The baths were made fashionable by visits from Queen Anne of Denmark, the wife of James I, in 1616 and by Charles II, who was seeking (in vain) a cure for Queen Catherine's infertility, and his court in 1663. They reached their heyday in the eighteenth and nineteenth century, helped by the efforts of Richard 'Beau' Nash who guided their development from 1702. His strict insistence on a moral code (banning nude bathing in 1737) helped elevate their previous reputation from little more than an excuse for holding orgies into an important stopping point for polite society.

Beautified with Georgian terraces and crescents and enriched by the thousands who came to take the waters, Bath was provided with all the latest amenities, receiving gas street lighting in 1819 and an electric power station in 1890. The town was badly bombed on 25 April 1942 in one of the 'Baedeker Raids' ordered against England's historic towns by Hitler in revenge for the RAF attack on Lübeck. The historic Assembly Rooms were destroyed and over 400 people killed, but Bath recovered its poise and continued to act as a magnet for visitors. By 2015, the Roman baths were attracting a million tourists a year, the descendants of their Roman ancestors who had deposited curse tablets and their Stuart forbears who had come in the hope of a little nude frivolity.

All this seems distant from the air, which privileges patterns and perspective over the minutiae of daily life. It is a means of map-making which opened up for the first time in the age of ballooning; the first British exponent was Thomas Baldwin who took to the air in 1785 and drew the earliest aerial views in his *Balloon Prospect about the Clouds*, in which he gushed about 'The endless variety of objects, minute, distinct and separate, tho' apparently on the same Plain of Level'. The first attempt at aerial photography was made by James Glaisher, head of the Meteorology department of the Greenwich Observatory in 1858 above Wolverhampton gasworks. Unfortunately, he rose far too fast, to well above 10,000 m (30,000 ft) and he and the crew nearly lost consciousness. Paralysed, his balloonist Henry Coxwell released the cord on the gas valve with his teeth and they descended safely. Britain's pioneer aerial photographer, therefore, only narrowly escaped the fate of Bath's mythical founder Bladud.

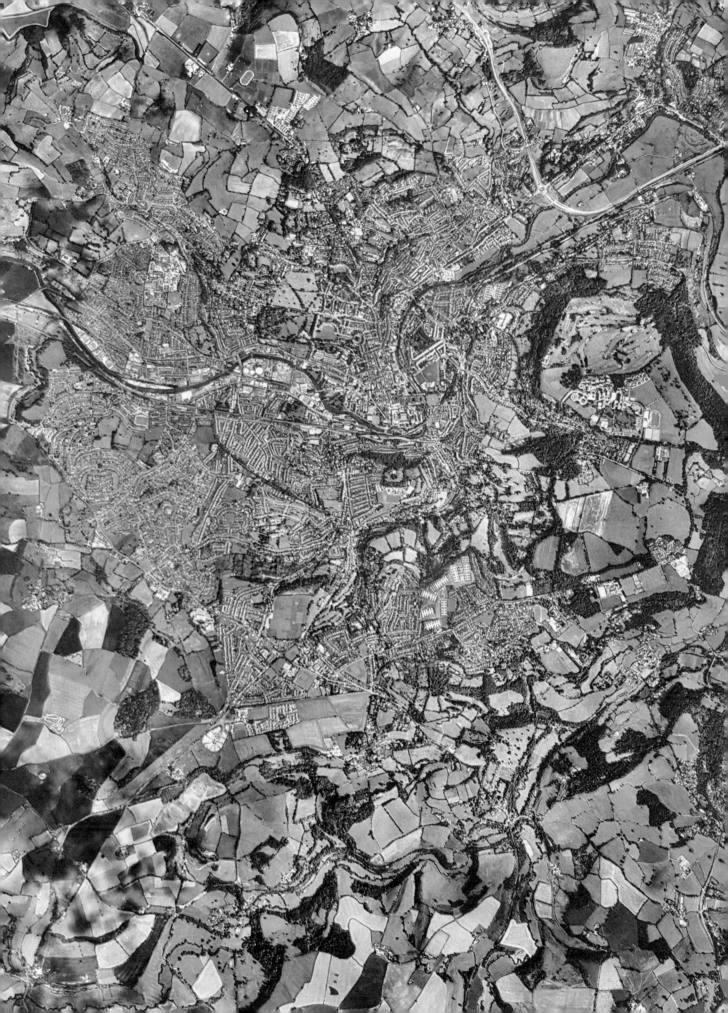

AMBLESIDE FLOOD MAP

The light blue shading on this 2014 Environment Agency map of Ambleside in the Lake District shows areas which were considered to be at risk of flooding between once in a century and once in a thousand years. Unfortunately floods on 5–6 December the following year inundated much of the town, suggesting the frequency of such flood events may need to be revised.

Britain is a low-lying island, with many coastal regions barely a metre (a few feet) above sea level. The sea which brought the nation economic riches and naval power just as often threatened the towns and villages which lay close to its shorelines. Particular regions, and often those with a strong maritime heritage, have always been prone to flooding.

As early as 1287, records exist of two enormous storms within a year which lashed the east and southeast coasts of England, destroying the town of Winchelsea and depriving New Romney of its port status, as the diversion of a river left it stranded inland. In 1362, a great storm lashed the Suffolk port of Dunwich, destroying much of it and leaving the rest submerged beneath the waves. It did not, though, deprive it of its parliamentary representation, and for the next 475 years it duly returned two MPs at each election, despite having virtually no inhabitants.

The Great Storm of 1703, which affected a great swathe of England, was one of the worst the country has endured. Hundreds of sailors were drowned and the Somerset Levels, amid whose marshes Alfred the Great had hidden from invading Danish Vikings in 878, experienced severe flooding. Daniel Defoe, the satirist and author of *Gulliver's Travels*, was moved to write a full-length account of the disaster in 1704. In *The Storm*, he relates that 123 people died in London and as many as 8,000 in total, including those who drowned on ships.

The modern age brought no relief, despite improvements in flood defences, including barrier walls and dikes. The collapse of the recently built Dale Dyke Dam in the Great Sheffield Flood of March 1864 led to further refinements, as it was found the dam's thickness was inadequate and overflow pipes had not diverted the flood. In 1953 another huge storm caused the North Sea Flood, in which over 300 people died along the coast of eastern England, as sea levels rose by up to 6 m (20 feet), while the ferry MV *Princess Victoria*, sailing from Stranraer in the west of Scotland to Larne in Northern Ireland, sank with the loss of 133 on board, in one of Britain's worst maritime disasters.

By 2015, the full panoply of scientific prediction and knowledge of flood prevention techniques still could not prevent catastrophic flooding. The Somerset Levels again suffered serious floods in 2013–14 and then, at the end of 2015 a series of particularly severe winter storms wrought havoc. Storm Desmond, on 5–6 December, dumped over 30 cm (1 ft) of rain on parts of Cumbria, causing bridges to collapse in Appleby, Keswick and Kendal, while Storm Eva on 24 December led to a repetition of the damage. Ambleside itself was flooded on the night of 5–6 December as the excess rains overwhelmed the banks of the River Rothay and caused the waters to come surging in to the town.

The enquiry into this and other floods forensically catalogued the damage and advised on measures which might be taken to reduce the chances of it occurring again. It showed those areas which might have expected such an event only once in a millennium, as though to suggest that such an event could not have been foreseen. Yet to anyone with a long view of history, looking back at the great floods that have stuck Britain over the past thousand years, the 2015–16 floods form just the latest in a long catalogue of such catastrophes.

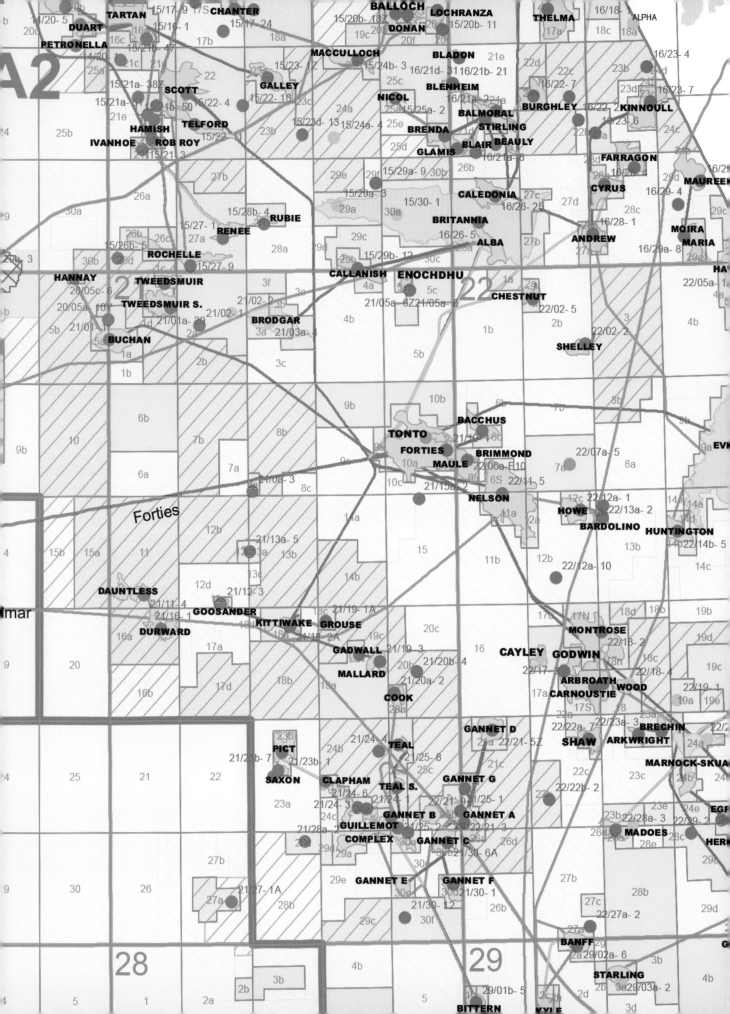

NORTH SEA OIL MAP

The complex mosaic of yellow blocks and cross-hatching in this 2016 map of the North Sea illustrates the development of an industry which provided an economic boon to 1970s' Britain. The Forties field, shown here, was the biggest of the North Sea oil fields whose royalties made the country an unexpected hydrocarbon superpower for a limited period.

Before the 1960s Britain's relationship with petrochemicals had been primarily that of an imperial power. Governments sought to ensure the security of oil supplies principally by calculated interventions in countries such as Iran, where the British were implicated in the 1953 coup which overthrew Mohammad Mossadegh, the radical prime minister who had tried to nationalize the Anglo-Iranian Oil Company.

There had been very limited oil discoveries in Britain; in 1851 it had been extracted from shale in West Lothian in Scotland but production made only a tiny dent in the thirst for oil which grew after the invention of the internal combustion engine in the 1870s. A single find in 1959 of natural gas in Groningen in the Netherlands changed all this. It became apparent that the geological structure of the North Sea meant that further discoveries were likely, although the prohibitive cost of drilling in offshore waters impeded initial exploration.

The first large find was of gas, struck in the West Sole field, off the Yorkshire coast, by British Petroleum's *Sea Gem* platform in September 1965 (though initial celebrations were dampened when the rig sank three months later with the loss of thirteen lives, in the North Sea's first major disaster). By 1969 oil had been struck in the Montrose field east of Aberdeen, followed soon after by the giant Forties field in 1970 and the Brent field a year later.

The system of licensing fees for exploration and royalties payable on gas and oil production provided an economic shot in the arm for a country which was struggling to restructure its traditional industries and facing increased competition from emerging industrial giants such as Japan. Production became even more profitable after the 1973 oil shock when the principal Middle Eastern oil-producing countries imposed oil embargoes in response to Western countries' policies which they said favoured Israel in its struggle with the Palestinians. As oil importers sought alternative sources of supply, and prices rose generally, North Sea oil found new customers.

Monthly oil production reached a peak of about 85 million barrels a month in early 1985 and annual production remained near these levels, until it began to tail off in 1999–2000. New oil fields opened up (such as the Buzzard, identified in 2001, with estimated reserves of 400 million barrels) as old ones dried up and the industry was generally considered an economic success story (though there were occasional set-backs such as the loss of 167 lives after an explosion on the *Piper Alpha* oil platform in 1988).

The revenues which accrued to the United Kingdom treasury (some £10.9 billion in 2011) became the source of political tension as nationalist politicians in Scotland argued that the income from North Sea oil should be repatriated, or that the country should become independent and take complete control of the royalties.

But by 2016 oil prices had tumbled, reaching a low of $28 a barrel for Brent crude (the North Sea's benchmark product), as opposed to $110 just eighteen months earlier. The British exchequer was now actually making a small loss from the oil industry, once tax breaks and investment rebates to the industry were taken into account. A sense of gloom spread over the North Sea and fears rose for the 400,000 jobs dependent on the oil industry. Having extracted some 40 billion barrels (and with perhaps around 25 billion barrels remaining), Britain's brief dalliance with the status of an oil exporting power was over. The North Sea is expected to provide significant continuing production for several decades yet, but as a cushion for Britain's economy and for its foreign policy, its time is over.

see more on next page >

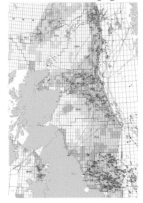

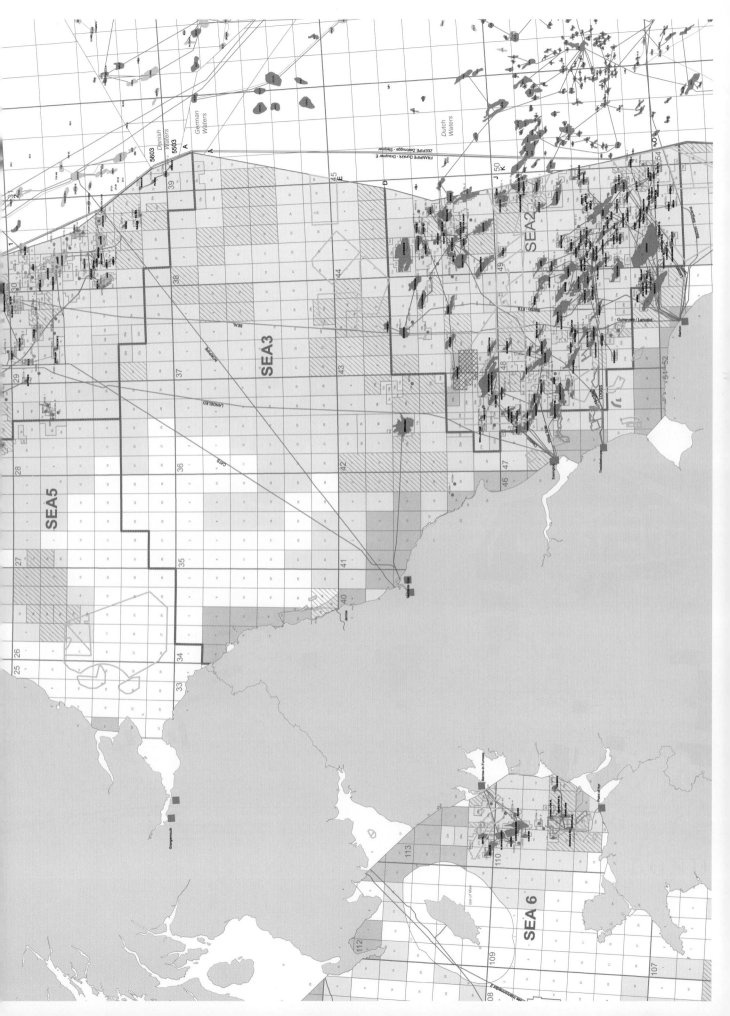

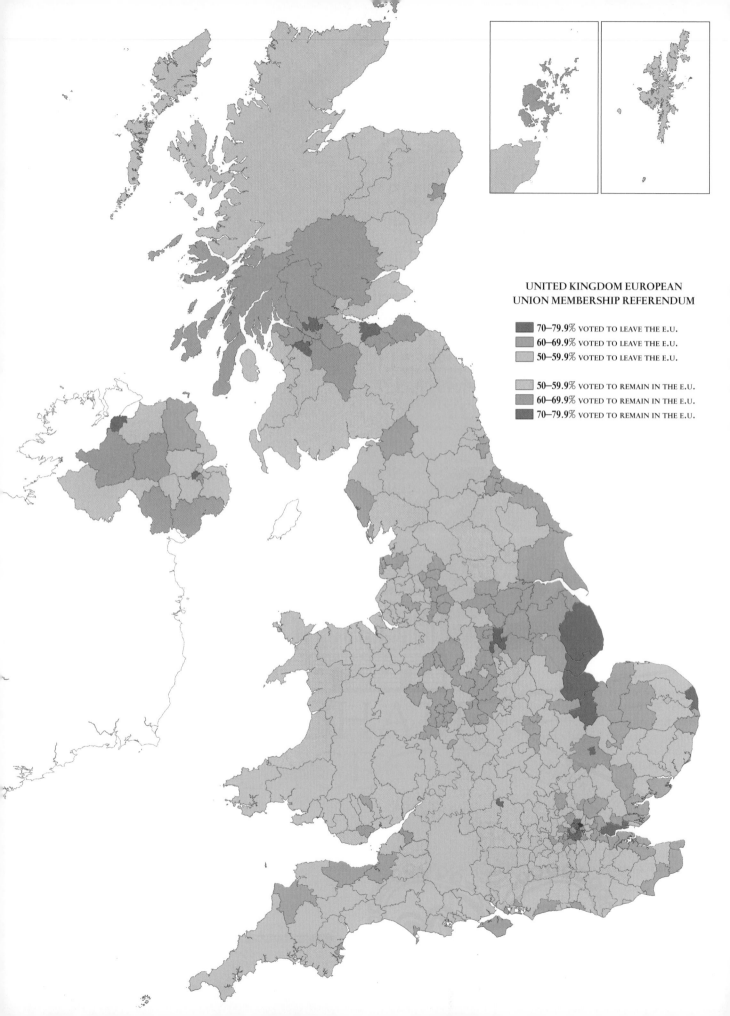

UNITED KINGDOM EUROPEAN
UNION MEMBERSHIP REFERENDUM

- 70–79.9% VOTED TO LEAVE THE E.U.
- 60–69.9% VOTED TO LEAVE THE E.U.
- 50–59.9% VOTED TO LEAVE THE E.U.

- 50–59.9% VOTED TO REMAIN IN THE E.U.
- 60–69.9% VOTED TO REMAIN IN THE E.U.
- 70–79.9% VOTED TO REMAIN IN THE E.U.

EUROPEAN UNION REFERENDUM VOTING PATTERN MAP

It is a map which marked a devastating blow to post-war Britain's political status quo. The shades of red spreading across the countryside – deepest in the east and northeast of England – mark those areas which voted to leave the European Union in the referendum held on 23 June 2016 in a vote whose closeness (at 52 to 48 per cent) belied its profound political consequences.

The European Union had at its heart the desire of the architects of the post-Second World War settlement to avoid any repetition of that catastrophic conflict by creating a system within which Germany and France could gain more by co-operation than they ever could by fighting. When the European Economic Community (the predecessor of the EU) was established by the Treaty of Rome in 1957, Britain stood aside, more concerned still with its colonial possessions and former colonies as it struggled to find a post-imperial role.

When Britain finally did apply to join the EEC in 1961, it was rebuffed, vetoed by Charles de Gaulle's France and it took a decade of negotiations before the British were finally admitted in 1973. There had never been national unanimity on the issue, and sceptics who championed British exceptionalism over European integration sought to leave the bloc from the very first. Such doubts were quelled, however, by a referendum in 1975 which confirmed Britain's EEC membership with a majority of 65 per cent voting to stay in.

British Eurosceptics, particularly within the Conservative Party, continued to campaign against Britain's membership, railing against what was seen as a leeching of power away from Westminster and towards the EEC's headquarters in Brussels. European institutions were strengthened by the Treaty of Maastricht in 1993 (which also changed the bloc's name to the European Union) and then further by the Treaty of Lisbon in 2007, which reduced the number of areas where individual nations could veto European Union decisions.

Britain had achieved opt-outs from several core EU policies, most notably from adopting the Euro, the single European currency, when it was launched in 2005, and it remained outside the Schengen area, within which holders of a single visa could travel freely without further checks. Yet this did not quell a growing movement for reassessing the country's relationship with the EU which coincided with the appearance of populist movements elsewhere (most notably France's Front National) that sought to promote a nationalist political agenda over moves to further European integration.

The backwash from the Arab Spring movement in North Africa and the Middle East, and particularly the outbreak of a devastating civil war in Syria after 2011, led to a large increase in refugees making their way to the EU overland through the Balkans and by sea across the Mediterranean. This fed into pre-existing concerns about a perceived dilution of British identity as a result of large numbers of EU citizens, mainly from eastern Europe, settling in Britain and fuelled calls for a further referendum on whether Britain should remain inside the EU.

Prime minister David Cameron, under pressure from the anti-EU United Kingdom Independence Party and large elements of his own Conservative party, acceded to these demands. His failure to extract significant concessions from other European leaders in negotiations early in 2016 and a campaign for Britain's exit (or Brexit) from the EU which appealed to voters in deprived areas by highlighting the increasing levels of immigration from Europe and the large financial gain which it was claimed Britain would derive from departing, resulted in a referendum result in favour of leaving the EU.

Large urban areas, particularly London, university towns such as Cambridge and Bristol, and Scotland and Northern Ireland, had, in contrast, voted to remain. The referendum revealed a nation split down the middle, in which, though the largest vote in British democratic history had gone in favour of leaving (17.4 million), equally the second largest ever vote had gone in favour of remaining (16.1 million). The government proceeded in any case with the process of leaving the EU, in March 2017 triggering Article 50 which began an irrevocable two-year process to negotiate the fine details of Brexit. Any politician who wanted to understand how divided the United Kingdom now was, however, needed only to look at the map of the referendum results.

Acknowledgements

The author wishes to thank the team at HarperCollins: Jethro Lennox, for commissioning the book, Keith Moore, Mimmi Rönning, Mark Steward, David White and Ewan Ross for their patient editorial work and Christopher Riches for his meticulous copy-editing.

While every effort has been made to trace the owner of copyright material reproduced herein and secure permission, the publishers would like to apologize for any omission and will be pleased to incorporate missing acknowledgements in any future edition of this book.

We are grateful to the following companies, organizations and individuals for supplying the historical maps included on these pages:

pp10–11 Cotton MS Nero D 1 f.187v Roman military roads, from 'Liber Additamentorum', c.1250-54 (vellum), Paris, Matthew (c.1200-59) / British Library, London, UK / Bridgeman Images

pp12–13 Caesar's Camp at St Pancras called the Brill (pen & ink with w/c on paper), Stukeley, William (1687-1765) / British Library, London, UK / Bridgeman Images

pp14–15 © Alnwick Castle / Northumberland Estates

pp16–17 Cotton MS Claudius D.VI f.10v 'Matthew Paris' Map of the Anglian Heptarchy (Anglo-Saxon Kingdoms)' c.1250 (vellum), English School, (13th century) / British Library, London, UK / Bridgeman Images

pp18–19 Cotton Tiberius B. V, Part 1, f.56v The Tiberius Map. Mappa Mundi (vellum), English School, (11th century) / British Library, London, UK / © British Library Board. All Rights Reserved / Bridgeman Images

pp20–23 Supplied by Philip Parker

pp24-25 Add MS 33991 f.26 Diagrammatic Map Of The British Isles 'Topographia Hiberniae', by Gerald of Wales (Giraldus Cambrensis) 1250 (ink & tempera on parchment), English School, (13th century) / British Library, London, UK / Bridgeman Images

pp26-27 Cotton Claudius D. VI, f.12v Map of Great Britain, illustration from 'Abbreviatio chronicorum Angliae', 1250-59 (vellum), Paris, Matthew (c.1200-59) / British Library, London, UK / © British Library Board. All Rights Reserved / Bridgeman Images

pp28-29 Part of an itinerary from London to Jerusalem, covering the journey from London to Beauvais, with representations of the main towns. / British Library, London, UK / © British Library Board. All Rights Reserved / Bridgeman Images

pp30-31 The Hereford Mappa Mundi Trust and the Dean and Chapter of Hereford Cathedral

pp32-35 Bibliotheque national de France

pp36-39 The Art Archives / Bodleian Libraries, The University of Oxford

pp40-41 Map of the British Isles, from 'Geographia' (vellum), Ptolemy (Claudius Ptolemaeus of Alexandria)(c.90-168)(after) / Biblioteca Marciana, Venice, Italy / Bridgeman Images

pp42-45 The National Archives, London UK.

pp46-47 By Lewis, John [Public domain], via Wikimedia Commons

pp48-49 The National Archives, London UK.

pp50-53 Lansdowne 204 ff.226-7 Map of Scotland by John Harding, c.1450 (ink & tempera on parchment), English School, (15th century) / British Library, London, UK / Bridgeman Images

pp54-55 © Bristol Records Office / Bristol Archives

pp56-57 Cotton Augustus I.i f.18 Plan of Brighton under Attack, 1539 (ink & tempera on parchment), English School, (16th century) / British Library, London, UK / Bridgeman Images

pp58-61 Plan of Dover / British Library, London, UK / © British Library Board. All Rights Reserved / Bridgeman Images

pp62-65 A map of the coast of Cornwall and Devon from the Scilly Isles and Land's End to Exeter, showing the coastal defences, 1539-40 (vellum), English School, (16th century) / British Library, London, UK / © British Library Board. All Rights Reserved / Bridgeman Images

pp66-67 Add MS 22721 f.10v Tidal Chart of England, c.1540-50 (ink & tempera on parchment), English School, (16th century) / British Library, London, UK / Bridgeman Images

pp68-69 Cotton Augustus I ii f.65 Map of the English Channel (English South Coast and French North Coast) by Jean Rotz, 1542-44 (ink & tempera on parchment), French School, (16th century) / British Library, London, UK / Bridgeman Images

pp70-73 By http://wellcomeimages.org/indexplus/obf_images/31/0d/56e289450563352508a80ade6f26.jpgGallery: http://wellcomeimages.org/indexplus/image/M0008080.html, CC BY 4.0, https://commons.wikimedia.org/w/index.php?curid=36333756

pp74-77 'The Englishe victore agaynste the Schottes by Muskelbroghe' 1547 (engraving), English School, (16th century) / British Library, London, UK / Bridgeman Images

pp78-81 and endpapers The National Archives, London UK.

pp82-83 The National Archives, London UK.

pp84-85 © Museum of London

pp86-87 Reproduced with permission from the National Library of Wales

pp88-89 Street Map of Cambridge / British Library, London, UK / © British Library Board. All Rights Reserved / Bridgeman Images

pp90-93 Map of the County of Cornwall, 1579 (hand-coloured engraving), Saxton, Christopher (c.1542-c.1610) / Private Collection / Bridgeman Images

pp94-95 The National Archives, London UK.

pp96-99 Beacons in Kent, 1585 (w/c on paper), Lambarde, William (fl.1585) / British Library, London, UK / © British Library Board. All Rights Reserved / Bridgeman Images

pp100-101 Map of the British Isles / British Library, London, UK / © British Library Board. All Rights Reserved / Bridgeman Images

pp102-105 Two maps: 1) The Sheldon Tapestry, showing detail of Warwick Castle, town and area, c.1590-1600 (tapestry), English School, (16th century) / Social History Collection, Warwickshire Museum Service / Bridgeman Images. 2) The Sheldon Tapestry, c.1590-1600 (tapestry), English School, (16th century) / Social History Collection, Warwickshire Museum Service / Bridgeman Images

pp106-109 A map of London / British Library, London, UK / © British Library Board. All Rights Reserved / Bridgeman Images

pp110-111 Reproduced with permission from Surrey History Centre

pp112-115 Reproduced by permission of the National Library of Scotland, Edinburgh, UK.

pp116-117 The kingdome of Irland. A map of Ireland, made in 1610. / British Library, London, UK / © British Library Board. All Rights Reserved / Bridgeman Images

pp118-119 Scotland / British Library, London, UK / © British Library Board. All Rights Reserved / Bridgeman Images

pp120-121 Historical Images Archive / Alamy Stock Photo

pp122-125 Panoramic view of part of the River Thames, known as the Visscher panorama. The map shows London as it would have been around the year 1600. The Globe Theatre can be seen in the foreground. / British Library, London, UK / © British Library Board. All Rights Reserved / Bridgeman Images

pp126-127 A Direction for the English Traviller, Distance Chart, 1635 (ink on vellum), Simmons, Matthew (17th Century) / British Library, London, UK / Bridgeman Images

pp128-131 Royal Collection Trust / © Her Majesty Queen Elizabeth II 2017

pp132-133 A Description of the seidge of Newarke upon Trent, with the fortifications about the Towne as also the forme of all the Entrenchements, forts, Redouts... / British Library, London, UK / © British Library Board. All Rights Reserved / Bridgeman Images

pp134-135 © National Maritime Museum, Greenwich, London

pp136-137 By Christopher Wren - This file was provided to Wikimedia Commons by Geographicus Rare Antique Maps, a specialist dealer in rare maps and other cartography of the 15th, 16th, 17th, 18th and 19th centuries, as part of a cooperation project. Public Domain, https://commons.wikimedia.org/w/index.php?curid=14682729

pp138-139 Chart of the Bristol Channel, 1671 (engraving), Seller, John (1632-97) / British Library, London, UK / Bridgeman Images

pp140-143 By John Ogilby [Public domain], via Wikimedia Commons

pp144-145 Map of Hertfordshire, 1676 (coloured engraving), Seller, John (1632-97) / British Library, London, UK / Bridgeman Images

pp146-147 A map of England and Wales with an alphabetical table of all the cities and market towns in England and Wales / British Library, London, UK / © British Library Board. All Rights Reserved / Bridgeman Images

pp148-149 Map of Postroads in the vicinity of London, 1713 (engraving), Willdey, George (fl.1707-37) / British Library, London, UK / Bridgeman Images

pp150-151 A map of the Grosvenor Estate. The boundaries of the St. George parish are outlined in red, with the properties in the estate in pink, and parks and open spaces in green. The plan shows the proposed new street plans for Grosvenor Square / British Library, London, UK / © British Library Board. All Rights Reserved / Bridgeman Images

pp152-155 A map of North Britain [Scotland] / British Library, London, UK / © British Library Board. All Rights Reserved / Bridgeman Images

pp156-159 A plan of the cities of London and Westminster. A scale of 26 inches to the mile / British Library, London, UK / © British Library Board. All Rights Reserved / Bridgeman Images

pp160-161 Map of Derbyshire, 1767 (engraving), Burdett, Peter Perez (1735-93) / British Library, London, UK / Bridgeman Images

pp162-163 A Plan of the River Tees, and of the intended navigable Canal from Stockton by Darlington to Winston in the Bishoprick of Durham. Surveyed by R. Whitworth, 1768 (coloured engraving), English School, (18th century) / British Library, London, UK / Bridgeman Images

pp164-165 Map of Lancashire, 1753 (coloured engraving), Bowen, Emanuel (fl.1777) / British Library, London, UK / Bridgeman Images

pp166-167 Reproduced by permission of the National Library of Scotland, Edinburgh, UK.

pp168-169 London Docks / British Library, London, UK / © British Library Board. All Rights Reserved / Bridgeman Images

pp170-171 By Ordnance Survey - http://www.ordnancesurvey.co.uk/oswebsite/images/userImages/misc/aboutus/history/mudge_map_kent_1801.jpg, Public Domain, https://commons.wikimedia.org/w/index.php?curid=12349024

pp172-173 Public Domain

pp174-177 © Science Photo Library

pp178-179 Ref No. Q/RDE/26, held at the Somerset Heritage Centre, reproduced with kind permission of the South West Heritage Trust

pp180-181 Reproduced by permission of the National Library of Scotland, Edinburgh, UK.

pp182-185 Two railway maps / British Library, London, UK / © British Library Board. All Rights Reserved / Bridgeman Images

pp186-189 The National Archives, London UK.

pp190-193 Reproduced by permission of the National Library of Scotland, Edinburgh, UK.

pp194-197 Getty/Science & Society Picture Library

pp198-199 By John Snow [Public domain], via Wikimedia Commons

pp200-201 © Times Newspapers